The Sustainable
Built Environment

THE SUSTAINABLE BUILT ENVIRONMENT

Technical, Managerial, Legal and Economic Aspects

Edited by

Begum Sertyesilisik

and

Ahmed Al-Shamma'a

First published 2016 by
PALGRAVE

Palgrave in the UK is an imprint of Macmillan Publishers Limited,
registered in England, company number 785998, of 4 Crinan Street,
London, N1 9XW.

Palgrave Macmillan in the US is a division of St Martin's Press LLC,
175 Fifth Avenue, New York, NY 10010.

Palgrave is a global imprint of the above companies and is represented
throughout the world.

Palgrave® and Macmillan® are registered trademarks in the United States,
the United Kingdom, Europe and other countries.

ISBN 978–0–230–31444–3 paperback

This book is printed on paper suitable for recycling and made from fully
managed and sustained forest sources. Logging, pulping and manufacturing
processes are expected to conform to the environmental regulations of the
country of origin.

A catalogue record for this book is available from the British Library.

Library of Congress Cataloging-in-Publication Data
Names: Sertyesilisik, Begum, editor. | Al Shamma'a, Ahmed, editor.
Title: The sustainable built environment : technical, managerial, legal and
    economic aspects / [edited by] Begum Sertyesilisik, Ahmed Al Shamma'a.
Description: New York : Palgrave Macmillan, 2016. | Includes bibliographical
    references and index.
Identifiers: LCCN 2015038154 | ISBN 9780230314443 (paperback)
Subjects: LCSH: Sustainable construction. | Sustainable buildings. | BISAC:
    TECHNOLOGY & ENGINEERING / Environmental / General. | TECHNOLOGY &
    ENGINEERING / General. | TECHNOLOGY & ENGINEERING / Industrial Technology.
Classification: LCC TH880 .S858 2016 | DDC 690.028/6—dc23
LC record available at http://lccn.loc.gov/2015038154

Printed and bound by CPI Group (UK) Ltd, Croydon, CR0 4YY

Contents

List of Figures

List of Tables

Notes on the Editors

Begum Sertyesilisik

Begum Sertyesilisik is Associate Professor in the Department of Architecture at Istanbul Technical University. She has two Master's degrees in the fields of Construction Project Management and Business Administration from Istanbul Technical University and a PhD in Construction Contracts from Middle East Technical University, Turkey. She specialises in sustainability, project management, construction project management and contract and dispute management. She is a member of the Chamber of Architects in Istanbul. She has written books, articles and proceedings in the project management field, is on the editorial board of various journals and acts as a scientific board member in international conferences.

Ahmed Al-Shamma'a

Ahmed Al-Shamma'a is Dean of the Faculty of Technology and Environment at Liverpool John Moores University. Ahmed's research covers a wide range of areas in applied industrial science, including advanced microwave technologies for chemical synthesis, renewable energies from waste including biodiesel, bioethanol and biobutanol, NDT wireless sensors for the construction, healthcare, automotive and aerospace industries, material processing, bespoke software solutions to monitor real-time energy levels in various industrial applications and near-zero carbon initiatives for the energy sectors. He is one of the EU Scientific Officers on Renewable Energies and has obtained various supported applied research projects, funded nationally and internationally by the EU, UK and USA Ministry of Defense, Carbon Trust, Technology Strategy Board and also directly funded by industry through various knowledge transfer partnerships. The success of the research has led to the establishment of three spin-out companies in Liverpool John Moores University. He has published over 250 peer reviewed scientific publications, holds 15 patents, has successfully supervised 32 PhD students and has co-ordinated over 30 research projects.

Notes on the Contributors

Amr Sourani

Dr. Amr Sourani is Senior Lecturer in Construction Management and Quantity Surveying. He worked briefly as a structural engineer for an engineering consultancy and as a site engineer for three years for a construction contracting company. He has a BSc in Civil Engineering (1998), an MSc in Engineering Project Management (with distinction) from the University of Manchester Institute of Science and Technology, which awarded him the MSc Course Prize in Engineering Project Management (2002) and a PhD from Loughborough University (2008). In 2007, Dr. Sourani joined Liverpool John Moores University as Lecturer in Construction Management. He has published papers at international refereed conferences and in journals and is also a referee for a number of these, including *Construction Management and Economics, Journal of Engineering, Design and Technology, Proceedings of ICE – Engineering Sustainability, Proceedings of ICE – Management, Procurement and Law* and *International Journal of Professional Education and Practice.* His main research and teaching interests are in the areas of construction management, sustainable construction and sustainable procurement, sustainable development, project planning and control, procurement strategies, corporate social responsibility and social aspects of construction.

Anupa Manewa

Dr. Anupa Manewa is Senior Lecturer in Quantity Surveying in the School of the Built Environment, Liverpool John Moores University. She is a chartered quantity surveyor of the Institute of Quantity Surveyors, Sri Lanka, and the Australian Institute of Quantity Surveyors. She obtained her bachelor's degree in Quantity Surveying and an MPhil in Construction Management from University of Moratuwa, Sri Lanka and a PhD from Loughborough University, UK. She has published over 30 research publications, many of which are related to economic sustainability. Anupa's current research interests focus on lifecycle costing, building information modelling, adaptable buildings and sustainable construction. She is a fellow member of the Higher Education Academy, UK. Her research and teaching interests are in the fields of whole life costing, building information modelling, adaptable buildings and sustainable construction.

Dr. Basak Guçyeter

Dr. Basak Guçyeter is Assistant Professor in the Department of Architecture at Eskisehir Osmangazi University in Turkey. Her main research and teaching interests include energy performance of buildings, energy performance simulation, effects of occupant behaviour on energy performance of buildings, building physics (heat, air and moisture transfer), as well as environmental assessment methods and sustainable architecture.

Bernd Kochendoerfer

Bernd Kochendoerfer is Professor of Construction Project Management at Berlin Technical University (Technische Universität Berlin). His research and teaching interests include construction project management, productivity improvement, cost estimation, construction contracts, procurement and sustainability in the built environment.

David Phipps

David Phipps is Professor of Environmental Technology, Built Environment and Sustainable Technology in the School of the Built Environment at Liverpool John Moores University. He has extensive experience working with water utilities on both domestic water efficiency and wastewater treatment. Previously, he worked as a consultant to a number of major companies on process and quality control. His teaching experience includes undergraduate and postgraduate chemistry and industrial chemistry, biotechnology and biochemistry.

Derek King

Derek King is Principal Lecturer in the School of the Built Environment at Liverpool John Moores University. He is Programme Leader for Building Services Engineering and Architectural Engineering. He is the Chair of the Merseyside and North Wales region of the Chartered Institution of Building Services Engineers (CIBSE). He teaches undergraduate and postgraduate levels in the following subjects: environmental analysis and simulation, mechanical building services engineering, engineering principles, environmental science and materials. His research interests include building services engineering education internationally and sustainability in public health engineering.

H. Murat Gunaydin

H. Murat Gunaydin is Professor in the Department of Architecture at Istanbul Technical University. Professor Gunaydin received his B.Arch. from Yıldız Technical University in 1991. He studied for his MSc in the area of project and construction management at Istanbul Technical University between the years 1991–1993. Then he received his PhD in construction engineering and management from Illinois Institute of Technology in 1999. Professor Gunaydin has worked as a faculty member and served in different managerial posts (i.e. as Chair of Department of Architecture, Vice Dean of the Faculty of Architecture, Vice Rector responsible for campus development/infrastructure projects/programmes and as the Dean of the Faculty of Architecture) at Izmir Institute of Technology. His main research and teaching interests include construction engineering and management, quality management in the construction industry, information technologies for project management, as well as intelligent buildings and energy efficient design.

Jack Rostron

Jack Rostron is Judicial Member of the Property Chamber First-tier Tribunal in England and Visiting Lecturer in Construction Management at Manchester University. He is a solicitor, town planner and surveyor specialising in Planning and Property Law in the High and County Courts of England.

Laurence Brady

Laurence Brady is Senior Lecturer in the School of the Built Environment at Liverpool John Moores University. His research and teaching interests include the technologies and management of building services engineering in areas of design, installation, operation and maintenance, renewables, sustainability and low- and zero-carbon technologies.

Margarete Roigk

Margarete Roigk has a PhD from the Berlin Technical University (Technische Universität Berlin). Her research focuses on sustainability and the sustainable built environment. She runs a company called "Architekturbüro Roigk" in Berlin.

Tofigh Tabesh

Tofigh Tabesh is a PhD candidate at Istanbul Technical University. His research and teaching interests include energy efficiency, life-cycle cost, courtyards and atriums, sustainability and the sustainable built environment.

Acknowledgements

We would like to thank our author team (Amr Sourani, Anupa Manewa, Bernd Kochendoerfer, Margarete Roigk, Basak Guçyeter, H. Murat Gunaydin, Tofigh Tabesh, Laurence Brady, David Phipps, Derek King and Jack Rostron) for their work towards accomplishing this project.

We are grateful to Helen Bugler, Jenny Hindley and Isabel Berwick from Palgrave for their encouragement throughout the preparation phase of this book.

Finally, we would like to thank CIBSE for their permission to reproduce Figure 6.7, adapted from CIBSE (2007), Solar Heating Design and Installation Guide.

Begum Sertyesilisik and Ahmed Al-Shamma'a

1 Introduction to the Sustainable Built Environment

Begum Sertyesilisik, Ahmed Al-Shamma'a, Amr Sourani, Anupa Manewa, Bernd Kochendoerfer, Margarete Roigk, Basak Guçyeter, H. Murat Gunaydin, Tofigh Tabesh, Laurence Brady, David Phipps, Derek King and Jack Rostron

The world's habitat is deteriorating in terms of global warming, the extinction of natural resources and the loss of biodiversity, due to the environmental footprint of human beings through the emission of greenhouse gasses arising from industrialisation and the exploitation of natural resources. The construction industry is one of the industries that affects the environment adversely. The construction industry affects the environment through its outputs (the built environment) and through production processes (material manufacturing, building processes). For this reason, there is a need for a sustainable built environment based on principles that allow the environmental footprint of the built environment to be reduced. A sustainable built environment aims to achieve zero negative impact on the environment.

Aims and rationale

This book aims to provide knowledge on the technical, managerial, legal and economic aspects of the sustainable built environment (Figure 1.1).

The book is designed to help researchers, academics and students as well as practitioners in the field of construction. There are nine chapters in the book, all written by expert contributors. The content of each is summarised here:

This first chapter provides information on the rationale and content of the book.

Chapter 2 focuses on sustainable procurement. There is a need to move from the adoption of "conventional" procurement strategies towards more "sustainability-oriented" procurement strategies. The chapter aims to explain the concept of "sustainable procurement" and its application to the context of construction. The chapter outlines concepts and principles relating to sustainability, and their relevance towards achieving organisational goals as well as their adaptation to the construction industry. Much can be done to achieve sustainability within the value-for-money approach, which is the aim of construction procurement. Integrating sustainability into the project brief and contract specifications, choosing an appropriate procurement system from a sustainability perspective, using multi-criteria decision-making techniques, evaluating and selecting contractors on the basis of value and provision of incentives and rewards are all examples of actions that can improve the contribution of procurement strategies towards sustainability. There are also other enablers that can enhance the development of "sustainability-oriented" procurement strategies. These enablers

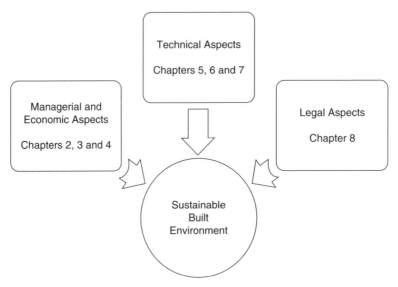

Figure 1.1 Structure and scope of the book

include knowledge and perception, and political and regulative, financial, instrumental, logistical, organisational and management and strategic enablers.

Chapter 3 focuses on cost modelling for sustainability. Sustainability is a term that encompasses a broad definition, which focuses on maintaining and improving the social, economic and environmental state of present and future generations. This broad scope of sustainability is usually rendered to quantifiable aspects that are directly related to reduction of the energy consumed or materials used/wasted. However, costs in the life cycle of buildings should be considered to be one of the essential parameters in assessment of sustainability measures, since any effort in achieving sustainable measures for the built environment is evaluated through financial value. Thus, in today's world, there is an absolute necessity to provide cost efficient applications for the establishment of the sustainable built environment. Chapter 3 focuses on determination and optimisation of life-cycle costs for sustainable construction and their relationship with energy conservation and management. It covers definitions such as cost, cost accounting and accounting for sustainability. It introduces methods and tools on how cost control systems are integrated into life-cycle considerations for the built environment. The chapter covers interdisciplinary methods for optimisation of the life-cycle costs, such as life-cycle engineering, facility management, primary energy and impact assessment, with an emphasis on their importance in decision making needed for the establishment of the sustainable built environment. Furthermore, the last section provides a brief exploration of current research on life-cycle cost assessment of buildings and their relationship with the sustainable built environment.

Chapter 4 continues the focus on managerial and economical aspects of the sustainable built environment. This chapter on the sustainable building process provides a detailed analysis of sustainable construction in three main phases, namely: pre-construction, construction and post-construction and outlines key aspects of each phase. The lean construction concept is investigated as a key for enhancing sustainability performance of the construction project management.

Chapter 5, which is on sustainable buildings, focuses on the definition of and drivers for sustainable buildings, building assessment tools and key technical aspects of sustainable buildings, including topics such as the selection of building materials, the building envelope, heating, ventilating and air conditioning.

Chapter 6 introduces low- and zero-carbon technologies used in sustainable buildings. Low- and zero-carbon technologies are becoming more relevant as awareness of environmental factors such as global warming increases. Low- and zero-carbon technologies can make the difference in creating a successful sustainable project. Nevertheless, the chapter points out that these technological solutions are only part of a strategy for minimising carbon emissions from buildings. In fact, intelligent architectural and structural design can utilise the building fabric and layout to absorb heating and cooling loads, as well as encouraging natural ventilation. The combination of renewable engineering services and effective environmental control by architectural means indicates that, to achieve low carbon targets for buildings, modern design teams must co-operate effectively.

Chapter 7 focuses on sustainability in utilities and water-efficient sustainable buildings, and aims to set out the broad principles of domestic water efficiency. Providing an adequate, secure, domestic water supply for intensely urbanised, developed societies is now a growing problem worldwide. Satisfying demand by increasing conventional supply is seen as unsustainable. Some limited help on the supply side may come from adopting alternative primary supplies such as rain- and stormwater capture or by employing grey- and blackwater for non-potable uses, though many technical problems still remain to be addressed. It is widely recognised that the only effective strategy will involve reducing demand by improving the efficiency of water use.

Chapter 8 focuses on the legal aspects of the sustainable built environment. It focuses on the interrelationship of sustainable development and environmental impact assessment. It defines sustainable development and the evolution of policy. It explains reporting requirements along with the indicators used as well as the technical process of assessment. It reviews the origins of planning controls, which essentially started in the early part of the twentieth century. The legal background to environmental impact assessment is explained. This includes the range of land uses and types of development that require an assessment to be produced before a planning application can be considered for approval. The process for the submission and consultation requirements is described, along with the environmental impact assessment regulatory requirements.

Chapter 9 concludes the book, drawing together the material from each of the different chapters.

The book is aimed at undergraduate and graduate students, researchers, tutors and practitioners in the field of construction. Readers can choose between reading the book from the beginning to the end or starting with the chapter in which they have most interest. We hope that the readers will enjoy reading this book and increasing their knowledge about the sustainable built environment.

2   Sustainable Procurement

Amr Sourani and Anupa Manewa

2.1 Introduction

This chapter aims to explain the concept of "sustainable procurement" and its application with regard to the construction context. The first section presents the concepts and principles related to sustainability and their relevance towards achieving organisational goals. Subsequent sections elaborate how these concepts and practices have been adopted in the construction industry. This chapter highlights the essential need for a renewed understanding of sustainable procurement in order to improve the performance of sustainability within the construction process.

2.2 Sustainable procurement

Procurement is considered a key function of many businesses and organisations. A robust "procurement system" plays a vital role in achieving the whole life value targets of the business. A good procurement system will improve efficiency and effectiveness of delivery and ensure better value for money. It also minimises the risk of corruption and positively impacts on the investment climate, non-discriminatory practices, transparency and accountability within a country (UN, 2006). Moreover, procurement plays an increasingly strategic role as a lever for sustainable development (Theron and Dowden, 2014).

Sustainable development has become a major focus for policies and research. It has been declared as "an overarching policy goal" by governments at the Earth Summit on Development and Environment (Parkin et al., 2003). The Sustainable Development Research Network (2002) describes sustainable development as "the most fundamental long-term challenge facing the world community". Given the increasing recognition of the concept, more than 200 definitions of sustainable development emerged (Parkin et al., 2003). Possibly the best known definition is the one introduced by The World Commission on Environment and Development (1987), which will be used in this chapter: "Development that meets the needs of the present without compromising the ability of future generations to meet their own needs." This definition is comprehensive to capture the different dimensions of sustainable development; in its broader sense, it comprises all the social, environmental and economic needs, as they must be in balance with each other to achieve sustainable outcomes in the long term (Brandon and Lombardi, 2005).

Considerations of "sustainability" are becoming more popular within the sustainability agenda goal. Organisations are compelled to integrate sustainability into their procurement processes and strategy. Therefore, "sustainable procurement" can be said to be one of the fast-growing fields of interest in corporate and government organisations across the world (McMurray et al., 2014). The World Summit on Sustainable Development (2002) defined that sustainable procurement is the process that promotes policies encouraging development and diffusion of environmentally sound goods and services.

The term "sustainable procurement" is multifaceted and is a mandatory criterion to be considered within the UK government's sustainability agenda. However, there is "no consistent definition in use across the public sectors that both policy makers and procurement professionals could relate to" (Sustainable Procurement Task Force, 2006). In response to this, the Sustainable Procurement Task Force (an organisation jointly funded by the Department for the Environment, Food and Rural Affairs and HM Treasury) devised what it called a "versatile" definition of sustainable procurement, encompassing the three dimensions of sustainable development (economic, social and environmental). They devised the definition of a sustainable procurement as follows:

> Sustainable Procurement is a process whereby organisations meet their needs for goods, services, works and utilities in a way that achieves value for money on a whole life basis in terms of generating benefits not only to the organisation, but also to society and the economy, whilst minimising damage to the environment (Sustainable Procurement Task Force, 2006: 10).

Literature reveals the importance of integrating sustainable components (social, environmental and economic) within the product and processes of procurement. In essence, "construction product" refers to any product or "kit" that is produced and placed on the market for permanent incorporation into construction works or parts thereof and the performance of which has an effect on the overall performance of the construction works with respect to the basic requirements for construction works (Construction Products Association, 2012). The underpinning activities (direct and indirect) that are required to produce such products are called processes. Environmentally preferable procurement ("green procurement") is highly acknowledged in both public and private procurement protocols. The term "green procurement" is "a process whereby public authorities seek to procure goods, services and works with a reduced environmental impact throughout their life cycle when compared to goods, services and works with the same primary function that would otherwise be procured" (Commission of the European Communities, 2008). The construction sector needs to improve its environmental performance and has therefore adopted sustainable procurement policy instruments (Uttam, 2014). Although sustainable procurement has been widely acknowledged in policy circles around the world, very little is known about the exact extent to which sustainable procurement policies and practices are embedded within the practice of procurement professionals globally. The next section explains the sustainable procurement considerations of the construction industry.

2.3 Sustainable construction procurement

Construction is significant to the economy. The contribution that the construction industry makes to the national economy, environment and to the overall sustainability agenda is significant. Its contribution to the Gross National Product (GNP) of the whole world is approximately 10% (Hillebranbdt, 2000). In Great Britain, its contribution to the Gross Value Addition (GVA) was approximately 6% in 2012 (Rhodes, 2013) and the same percentage was maintained in 2014 (Department of Business, Innovation and Skills, 2014). Harvey and Ashworth

(1997) highlighted that the industry also represents over half of Britain's fixed capital investment and contributed to 6.6% of employment in 2013 (Rhodes, 2013). Such evidence demonstrates the significant role that the construction industry plays in a country's economy. Sustainable considerations are given great prominence within the construction project and process protocols, which help to achieve better value for a client's investment.

With reference to the Joint Contracts Tribunal (2014), "Construction Procurement" describes the activities undertaken by a client or employer who is seeking to bring about the construction or refurbishment of a building. The procurement process initiates through devising a project strategy, which entails weighing up the benefits, risks and budget constraints of a project to determine what the most appropriate procurement method is and what contractual arrangements will be required (JCT, 2014). Sustainability considerations are increasingly important and need to be integrated into construction procurement so that "sustainable development" can be achieved. The policy paper "Construction 2025: industrial strategy for construction – government and industry in partnership", published in 2013, explains that by 2025 the UK government needs to reduce the initial capital cost of construction and the whole-life cost of built assets by 33%, construction project duration by 50% and emissions by 50%. Construction procurement plays a vital role in achieving these targets. Moreover, the report "Accelerating Change" recommended that the construction industry must take responsibility for the sustainability of its products as well as its processes (Strategic Forum for Construction, 2002). In 2007, the government published the Sustainable Procurement Action Plan. Among the goals set out in the plan was for the UK to be "among the European Union (EU) leaders in sustainable procurement by 2009" and to achieve "a low carbon, more resource efficient public sector" (DEFRA, 2007).

Sustainable construction advocates the application of sustainable development principles within the construction industry. In its broader sense, "sustainable construction is the set of processes by which a profitable and competitive industry delivers built assets (buildings, structures, supporting infrastructure and their immediate surroundings) which: enhance quality of life and offer customer satisfaction; offer flexibility and the potential to cater for user changes in the future; provide and support desirable natural and social environments; and maximise the efficient use of resources" (Raynsford, 2000). Raynsford's definition puts emphasis not only on the product but also on the process. The definition introduces some aspects of social sustainability such as customer satisfaction and support for desirable social environments. It also introduces some aspects of environmental and economic sustainability such as maximising the efficient use of resources and emphasising the profitability and competitiveness of the industry. However, the definition does not fully capture all aspects of sustainable construction. A more comprehensive definition has been offered by Constructing Excellence (an organisation charged with driving the change agenda in construction), which introduces sustainable construction as the application of sustainable development in the construction industry and suggests that sustainable development is:

> all about ensuring a better quality of life for everyone, now and for generations to come, through: social progress which recognises the needs of

everyone; maintenance of high and stable levels of economic growth and employment, whilst; protecting, and if possible enhancing, the environment, and; using natural resources prudently (Constructing Excellence, 2004a).

Despite the variances among different definitions of sustainability, there is a wide acceptance that sustainable development integrates at least three dimensions: social, economic and environmental. Sustainable construction is about achieving a balance between these aspects of construction so that the costs and the benefits, in social, economic and environmental terms, are optimised. Criteria underpinning each of these dimensions are listed below (Hill and Bowen, 1997; GCCP, 2000; Addis and Talbot, 2001; Rethinking Construction's Respect for People Working Group, 2002; Rethinking Construction, 2003; TCPA and WWF, 2003; IDeA, 2003; ODPM, 2003; Constructing Excellence, 2004b; OGC, 2004; HM Government, 2005; National Audit Office, 2005a; OGC, 2005):

a) Social sustainability, underpinned by criteria such as: health and safety; stakeholders' involvement; training and development of the workforce; equality and diversity in the workplace; workforce conditions; user needs and satisfaction; employment creation.
b) Economic sustainability, underpinned by criteria such as: whole life costing; supporting local economies; fitness for purpose; economic Key Performance Indicators (KPIs); waste minimisation and management.
c) Environmental sustainability, underpinned by criteria such as: reducing energy consumption; reducing water consumption; specifying low environmental impact materials; reusing existing built assets; considering the use of renewable resources; minimising water, land and air pollution and preserving and enhancing biodiversity.

The environmental dimension has traditionally been the major focus of the literature on sustainability. In addition some publications in the literature have mentioned other dimensions of sustainability such as technical sustainability (Hill and Bowen, 1997; Ashley et al., 2003), cultural sustainability (Ofori, 1998; CIB, 1999; Langford et al., 1999), community sustainability (Ofori, 1998) and managerial sustainability (Ofori, 1998). In the context of the UK construction industry the triple bottom line concept, which focuses on social, economic and environmental sustainability, remains dominant.

Construction in the public sector includes a wide range of activities comprising major infrastructure and civil engineering projects as well as major building programmes (such as hospitals, schools, prisons and social housing), in addition to refurbishment and maintenance activities. While all government bodies are involved in construction activities, the involvement of these bodies can range from engaging in construction as a core business for some bodies (as in the case of the Highways Agency) to the occasional involvement of other bodies in significant construction projects (e.g. Olympics projects). Most government bodies, however, undertake repair and maintenance programmes (National Audit Office 2005a). Taking into account the significance, contribution and size of the public sector, considerable social, economic and environmental benefits can be gained from integrating sustainability into public procurement of construction projects (Scottish Parliament, 2013; National Statistics, 2014).

The urgent need to address sustainability principles in construction procurement has been increasingly acknowledged by construction professionals and academics. With reference to Rowlinson et al. (2000), sustainability is one of the developing themes in the context of construction procurement that is expected to grow significantly. Adetunji et al. (2003) pointed out that "client procurement policy" is one of the three highest ranked drivers for implementing sustainability. Official reports published in the UK have reflected the increasing emphasis on sustainable procurement. As mentioned earlier in the chapter, the report "Accelerating Change" recommended that the industry must take responsibility for the sustainability of its products as well as its processes (Strategic Forum for Construction, 2002). The DETR (2000) reported that a new programme would require all departments and agencies to adopt an action plan for more sustainable construction procurement. According to the Rethinking Construction's Respect for People Working Group (2002), addressing procurement in a sustainable way is a need that clients are starting to acknowledge.

Moreover, the Sustainable Procurement Action Plan (2007) described the targets in detail and specified how the government will achieve them while working towards a sustainable built environment. In June 2008, a joint industry-government strategy for sustainable construction was launched (HM Government and Strategic Forum for Construction, 2008). The strategy has since been agreed across government and covers both buildings and infrastructure (CIRIA, 2008). The strategy aims to provide clarity around the existing policy framework and the range of commitments, targets and actions relevant to sustainable construction (HM Government and Strategic Forum for Construction, 2008; CIRIA, 2008). A set of overarching targets has been presented to deliver the strategy. These targets are related to both the results of sustainable construction (which relate directly to sustainability issues, e.g. biodiversity) and the means of sustainable construction (i.e. the processes helping to achieve the end results). Among these are six targets representing the "ends" (climate change mitigation, climate change adaptation, water, biodiversity, waste and materials) and five targets representing the "means" (procurement, design, innovation, people and better regulation).

According to this governmental strategy, the overarching target of procurement is to "achieve improved whole life value through the promotion of best practice construction procurement and supply side integration, by encouraging the adoption of the Construction Commitments … in both the public and private sectors and throughout the supply chain" (HM Government and Strategic Forum for Construction, 2008: 7). The commitments referred to by the strategy relate to:

- ethical resourcing (enabling best value to be achieved and encouraging integration of the supply chain)
- valuing people (leading to a more productive workforce and facilitating recruitment and retention of staff and engagement of local communities)
- client leadership
- sustainability
- design quality (ensuring that the design is creative, imaginative, sustainable and capable of meeting delivery objectives and the needs of all stakeholders) and
- health and safety.

The stage of developing a procurement strategy is crucial in achieving sustainable procurement. There is a need to move from the adoption of "conventional" procurement strategies toward more "sustainability-oriented" procurement strategies. The procurement strategy:

> ... identifies the best way of achieving the objectives of the project and value for money, taking account of the risks and constraints, leading to decisions about the funding mechanism and asset ownership for the project. The aim of a procurement strategy is to achieve the optimum balance of risk, control and funding for a particular project (Office of Government Commerce, 2003).

In the context of public procurement in the UK, key tasks in developing a procurement strategy include: producing an outline business case, determining the procurement route (including contract strategy), producing output-based specification and criteria for selection and award and placing an advertisement in the Official Journal of the European Union (OJEU) if required (Office of Government Commerce, 2003).

The importance of considering sustainability at the stage of developing a procurement strategy has been highlighted in several publications. For example, the Improvement and Development Agency (IDeA) shows that "the key opportunity to consider environmental and social issues is at the earliest stages of the procurement cycle: identifying needs and building them into the design or specification. Adverse impacts should be managed out at this point" (IDeA, 2003: 3). The National Procurement Strategy for Local Government, in establishing how to achieve community benefits through procurement, provides the following recommendation:

> ... implement sustainable design and sustainable procurement strategies and build sustainability into procurement processes and contracts, where relevant to contract. Sustainability in design (buildings, infrastructure, urban, green spaces and products) and procurement should be addressed in risk-based strategies that complement the corporate procurement strategy and the community plan. Include environmental requirements in the user needs and specification at the earliest stages of the procurement process (ODPM, 2003).

OGC shows that the project brief, as part of the procurement process, must highlight the importance of sustainability and that the client must include sustainable performance objectives in the specification to enable tenderers to respond to these objectives (OGC, 2005). The following subsections explain how the aforementioned sustainability considerations could be integrated into a typical construction project.

2.3.1 Integrate sustainability in project brief and contract specifications

The project brief describes the intended project, specifies the expected outcome, outlines the role of contractor and highlights constraints and difficulties. Integrating sustainability principles into a proper project brief and clearly stating the multidimensional nature of sustainability will present sustainability dimensions in a way that cannot be ignored at any stage of project delivery. Relevant sustainability criteria can also be integrated into contract specifications, an issue that is highlighted by The Government Construction Clients' Panel (2000). However, in an environment dominated by "economic" performance measures, the issue of structuring sustainable construction contracts needs a "gradual" and "creative" approach, as argued by Gordon (1996). Gordon points out that structuring an innovative construction contract is a matter of replacing traditional fiscal indicators with others that have a true basis in environmental and social values. This does not mean that the economic or financial measures are no longer relevant. Instead, the underlying principle should be that indicators related to all the dimensions of sustainability should be incorporated in the construction contract in a balanced way.

Choosing an appropriate procurement system from a sustainability perspective

In considering the selection of a procurement system there are many noteworthy factors to be considered, which are associated with client needs, contractor requirements and project characteristics (Ambrose and Tucker, 2000). According to Newcombe (2000), the selection of an appropriate procurement path is "not as obvious as it sometimes appears and divergent choices are sometimes argued". What exacerbates the problem from a sustainability point of view is that despite the increasing recognition that sustainability criteria need to be accommodated in procurement strategies, such criteria have typically not featured as a key aspect in the selection of procurement systems. However, there are indications that the situation has undergone some change in some publications. For example, the criteria "control over sustainability issues" is now among the evaluation criteria of procurement systems, as appears in the evaluation template developed by the Office of Government Commerce (2003). Such systems have shown different levels of performance in attaining certain objectives such as speed, cost, certainty, flexibility, non-adversarial focus and risk allocation (Love et al., 1998; Alhazmi and McCaffer, 2000; Ambrose and Tucker, 2000). The challenge now is to assess the potential of these systems in attaining the criteria underpinning the different dimensions of sustainability. Following such assessment, informed decisions that consider the different sustainability criteria, the priority given to meeting such criteria and the performance of the different systems in attaining them can be made.

Using multi-criteria decision-making techniques

An assessment of sustainability needs to take into account the different criteria underpinning it, which might sometimes be in conflict. This could lead to the use of multi-criteria decision-making techniques, such as the Analytic Hierarchy Process (AHP). This process is a structured tool to aid complex decision making. It helps the user to establish priorities and ultimately make the best decision (Saaty, 1987). It works by breaking complex decisions down into a series of pairwise comparisons and then synthesising the results, helping to show both the subjective and objective aspects of the decision. The AHP also includes a helpful technique for determining a decision's consistency. Such techniques have been used for solving problems in construction management (Cheng and Li, 2002) and in selecting the construction procurement strategy (Al-Tabtabai, 2002). In some cases multi-criteria decision-making techniques were used where sustainability provided the basis for choosing the criteria for decision-making. Al-Tabtabai (2002) suggested that the use of AHP could be a powerful method for selecting the best procurement strategy, although sustainability is not presented as a major criterion in Al-Tabtabai's model. Cheung et al. (2001) also suggested the use of AHP in procurement selection, and there is a potential for using such techniques to make informed decisions that consider sustainability criteria in problems such as the selection of a contractor as well. The methods suggested by Al-Tabtabai (2002) or by Cheung et al. (2001) could be further expanded so that the criteria of the different dimensions of sustainability are incorporated.

Evaluating and selecting contractors on the basis of value

Following the Latham report (1994) entitled "Constructing the team" and the Egan report (1998) on "Rethinking Construction", the selection of consultants had tended to move away from appointments based on price alone towards selection procedures that are based on balancing quality and price (Addis and Talbot, 2001). Sustainability values are becoming increasingly important within the value-for-money approach which allows for the exclusion of contractors who have committed serious misconduct, on issues such as health and safety, from tendering (GCCP, 2000). One action that can bring sustainability values to selection procedures is requiring the contractors to register under the Considerate Contractor Scheme which helps them to act in a socially responsible way (Addis and Talbot, 2001).

Provision of incentives and rewards

The role of incentives as a key initiative in introducing a move towards more sustainable construction was appreciated by CIB (1999). Incentives, taking the form of financial gain or improved contract terms, determine future behaviour. Rewards recognise past performance and usually take the form of improved access to new work or improved contract terms (Kenley et al., 2000). Both are means, within a strategic procurement framework, by which change can be driven in the industry (Kenley et al., 2000). One possible way of providing incentives to obtain more sustainability-oriented procurement is highlighted by Casella Stanger et al, a company which offers sustainability consultancy services (2002). At the tender stage, contractors may be encouraged to identify sustainable solutions that can result in life-cycle savings.

The actions described here could enhance the development of "sustainability-oriented" procurement strategies. Other actions that could enhance the development of "sustainability-oriented" procurement strategies include:

- knowledge and perception issues (such as developing a common understanding of what constitutes sustainable development)
- political and regulatory issues (such as introducing more mandatory influence)
- financial issues (such as availability of funding)
- contractual issues (such as emphasising the importance of sustainability in tender evaluation and selection procedures)
- instrumental issues (such as utilisation/enhancement of existing assessment and measurement techniques and tools to consider sustainability)

- logistical issues (such as allowing sufficient time in the programme to address and assess sustainability issues)
- organisational and management issues (such as facilitating publicity of actions taken by public procurers towards addressing the sustainability agenda) and
- strategic issues (such as promoting Corporate Social Responsibility policy and implementation).

2.4 Sustainable procurement practices adopted in different countries

Different countries have adopted different approaches to make their procurement process a sustainable one. The majority of those approaches are focused on improving economic, environmental and social well-being; however, minor changes could be identified in their priorities, which depend on their national/local sustainable procurement policy guidance.

Few studies have been undertaken to identify the sustainable procurement practices adopted in different countries. Having analysed the sustainable procurement guidance of Organisation for Economic Co-operation and Development (OECD) of 30 countries, Walker et al. (2014) identified that the majority of countries are highly focused on economic aspects (e.g. competitive tender, market based, innovation, cost reduction, etc.) within their procurement policy when compared to environmental (e.g. product criteria, environment performance, eco-labelling, life cycle best practice, reduce-recycle-reuse etc.) and social aspects (equality, ethics, fair trade, human rights etc.). Having undertaken a comparative study on sustainable procurement practices adopted in Australian and Malaysian contexts, Islam and Siwar (2013) underlined a significant difference between those two practices. In particular, Australian public organisations placed stronger emphasis on safety aspects of sustainable procurement while Malaysia placed greater importance on diversity. Moreover, the findings from a recent study undertaken within the Sri Lankan context explain that the country has begun to pay more attention to integrating environmentally sustainable criteria within its public procurement policies/guidelines. For example, their procurement policy encourages the industry to procure environmentally friendly materials, green products, renewable fuels and also promotes carbon credits for renewable energy projects (Peiris, 2013).

Japan is considered to be one of the leading countries in relation to green public procurement which might be attributed to the strength of policy leadership at the top level in both the public and private sectors, resulting in a comprehensive and integrated approach (Ochoa et al., 2003; Thomson and Jackson, 2007). Examples include the development of eco-labelling, purchasing guidelines, product lists, economic instruments, a mandatory reporting system, life cycle analysis information and a public awareness programme (Thomson and Jackson, 2007).

Germany is another example of a country's high-level political commitment, which focuses on close co-operation and sharing of good practices between the federal and regional levels (Alejandre, 2010). A mandatory target for all authorities at the federal level is to use life-cycle costing to ensure energy efficient and environmentally friendly public procurement. German regulation of procurement requires public procurers to demand an analysis of minimised life-cycle costing in tenders (Alejandre, 2010). At the federal level, wood products procured by federal administrations must be obtained from sustainable sources. However, imposing mandatory requirements with regard to sustainable procurement has been questioned by Thomson and Jackson (2007) who argue that, in general, requiring organisations to develop green procurement policies (an approach taken by Japan) has been more successful than legislating that organisations should consider environmental impacts (as in the case of Germany).

The Hong Kong context has been examined by Liu et al. (2012), who argued that although the sustainability concept is supported, it is not fully implemented in Hong Kong in comparison to other countries such as the UK. For example, despite the emphasis and the attention given to green issues and sustainability in Hong Kong, very few organisations in the construction industry obtained ISO14001 certification. Furthermore, the dominance of traditional procurement arrangements in Hong Kong may not be helpful in delivering sustainability; such arrangements are characterised by poor constructability (leading to waste and longer project durations), variations (leading to re-work, delays and inefficient ordering and operations) and competitive tendering (fostering individualistic and opportunistic behaviour through claims and disputes) (Liu et al., 2012).

Table 2.1 summarises the most frequent sustainable procurement practices adopted in different countries, numbered one to five in order of priority.

Table 2.1 Sustainable procurement practices adopted in different countries

	Sustainable procurement practice	UK	Western Europe	USA/Canada	Australia	Malaysia
1	Purchases from local suppliers	1	1	1	2	1
2	Purchases from small suppliers	2	2	2		4
3	Ensures the safe, incoming movement of a product to an organisation's facilities	3	4	4	1	2
4	Ensures that suppliers comply with child labour laws	4				5
5	Ensures that suppliers' locations are operated in a safe manner	5			5	3
6	Reduces packaging materials			3	3	
7	Asks suppliers to commit to waste reduction goals			5	4	
8	Purchases from MWBE (Minority and Women-owned Business Enterprises) suppliers				3	
9	Has a formal MWBE supplier purchase programme				5	

Source: Adapted from Islam and Siwar (2013), Brammer and Walker (2011)

The findings (Table 2.1) indicate that the majority of the selected countries consider that "purchases from local suppliers" and "purchases from small suppliers" promote sustainable procurement to a greater extent. Moreover, Walker et al. (2014) emphasised that those countries that have not yet developed sustainable procurement guidance, such as developing countries or countries with economies in transition, should pay attention to the following economic, environmental and social considerations in their procurement policy:

- Economic – competitive tendering, market-based decision making, cost reduction and supporting innovation;
- Environmental – environmental product criteria, assessing environmental performance of suppliers, eco-labelling and life cycle analysis; and
- Social – supporting marginalised people, promoting equality and ensuring decent working conditions for suppliers' employees.

The literature reveals the importance of adopting sustainable procurement practices within the policy level of any organisation/country. Continuous assessment is required to monitor the procurement process and to identify whether it has considered the elements (social, economic, environmental) of sustainability. The introduction of a reward/penalty system is also required to establish sustainable procurement practices successfully. The next section explains the key challenges/barriers to sustainability.

2.5 Challenges and drivers for sustainable procurement

Although the concepts and the importance of sustainability and sustainable procurement are well recognised, their implementation has been far from easy. The main barriers to sustainable public procurement in central government, as identified by the National Audit Office (2005b), involve:

- conflict between sustainable procurement and reducing costs
- lack of leadership on sustainable procurement
- lack of integration of sustainable procurement into standard procurement processes, with the risk that sustainability issues will not be addressed
- lack of central control over procurement in departments, increasing the difficulty of enforcing sustainable procurement
- lack of knowledge about what sustainable procurement is and how to achieve it.

In addition to the challenges noted above and within the Sri Lankan context, Peiris (2013) identified "resistance to change", "lack of rewards/incentives for green products/processes" and "malpractices" as further key barriers.

Sourani and Sohail (2011) investigated the difficulties of achieving sustainable procurement in the UK. Their research identified the barriers that clients face when addressing sustainable construction in procurement strategies, as well as the parties most capable of removing those barriers. They interviewed sustainability experts from a wide range of professional and public sector organisations and identified 12 main barriers relating to: funding shortages and restrictions; a lack of understanding and available information; inadequate and inconsistent policies and regulations; insufficient and unclear advice and tools; vague definitions and varied interpretations of them; a split between capital and operational budgets; lack of time; a focus on short-term issues and considerations; the widespread assumption that sustainable procurement results in increased capital costs; a lack of enthusiasm for change; fragmentation in the construction industry and a lack of sufficient research and development. Sourani and Sohail concluded that to move the sustainable procurement agenda forward, several different parties need to be involved (Sourani and Sohail, 2011).

At an organisational level, Mohan (2010) identified the following significant drivers for achieving sustainable procurement:

- effective, organisation-wide policies to ensure that everyone is aware of the strategy
- training and guidance to help those involved in procurement to understand sustainable procurement and whole life costing
- regular audits and monitoring to assess where the organisation is in the context of sustainable procurement
- supporting and educating suppliers/linking up with other organisations to learn from their experience and
- pooling procurement through establishment of procurement consortiums where relevant.

2.6 Conclusions

In this chapter, the importance of sustainable procurement within the construction context has been analysed. The findings explain that "sustainable procurement" is one significant criterion in the procurement channel that leads to "sustainable construction" and "sustainable developments". Moreover, it has been identified that though different countries pay attention to different areas of sustainable procurement (health and safety, diversity etc.), their ultimate aim is highly targeted towards sustainable goals. The following factors should be addressed in order to move the sustainability agenda forward.

Firstly, the role of the government and other stakeholders needs to be revisited. For example, parties such as the government (including regulatory bodies) are best placed to deal with financial, regulatory, policy and guidance obstacles. Individual public procurers should provide adequate training, sufficient time and appropriate communication. Professional and educational bodies should raise the level of awareness of sustainable development throughout society. The supply chain should move towards further integration and users should stimulate demand for sustainable products.

Secondly, policy level requirements need to be more effectively integrated into buildings and a more explicit and practical understanding of their implications for the built environment needs to be developed. Moreover, this should be used to inform built environment decision-making processes in order to achieve built environments that are more sustainable.

Finally, there is a need for a renewed understanding of sustainable procurement in order to improve the performance of the construction product/processes.

The next chapter focuses upon cost modelling for sustainability and life-cycle assessment.

References

Assessment and evaluation

Al-Tabtabai, H.M. (2002). Construction Procurement Strategy Selection Using Analytical Hierarchy Process. Journal of Construction Procurement, 8(2), 117–132.

Brandon, P.S. and Lombardi, P. (2005). Evaluating Sustainable Development in the Built Environment. Blackwell Science Ltd, UK.

Cheng, E.W.L. and Li, H. (2002). Analytic Hierarchy Process: A Decision-Making Method for Construction Management. Proceedings of the First International Conference on Construction in the 21st Century (CITC 2002): Challenges and Opportunities in Management and Technology. Miami, FL, USA, 25–26 April, 2002, pp. 135–142.

Saaty, R.W. (1987). The Analytic Hierarchy Process – What It Is and How It Is Used. Mathematical Modelling, 9(3–5), 161–176.

Policies/Action plan

DEFRA (2007). UK Government Sustainable Procurement Action Plan – Incorporating the Government Response to the Report of the Sustainable Procurement Task Force. DEFRA, London.

ODPM (2003). National Procurement Strategy for Local Government. Office of the Deputy Prime Minister.

Office for National Statistics (2014). (Available at: http://www.ons.gov.uk/ons/taxonomy/index.html?nscl=Public+Sector+Employment. Accessed 29 June 2015).

Office of Government Commerce (2003). Procurement Guide 06: Procurement and Contract Strategies. London.

Office of Government Commerce (2004). Procurement Guide 10: Health and Safety. London.

Office of Government Commerce (2005). Procurement Guide 11: Sustainability. London.

Scottish Parliament (2013). Financial Scrutiny Unit Briefing – The Size of the Public Sector. (Available at: http://www.scottish.parliament.uk/ResearchBriefingsAndFactsheets/S4/SB_13-36.pdf. Accessed 29 June 2015).

The Government Construction Clients' Panel (June 2000). Constructing the Best Government Client: Achieving Sustainability in Construction Procurement – Sustainability Action Plan. Produced by the Government Construction Clients' Panel and Endorsed by the Office of Government Commerce.

Procurement

Addis, B. and Talbot, R. (2001). Sustainable Construction Procurement: A Guide to Delivering Environmentally Responsible Projects. CIRIA C571. CIRIA, London.

Alejandre. E. (2010). Best Practice on Green or Sustainable Public Procurement and New Guidelines. VISESA.

Alhazmi, T. and McCaffer, R. (2000). Project Procurement System Selection Model. Journal of Construction Engineering and Management, 126(3), 176–184.

Ambrose, M.D. and Tucker, S. N. (2000). Procurement System Evaluation for the Construction Industry. Journal of Construction Procurement, 6(2), 121–134.

Brammer, S. and Walker, H. (2011). Sustainable Procurement in the Public Sector: An International Comparative Study. International Journal of Operations and Production Management, 31(4), 452–476.

Commission of the European Communities (2008). Communication from the Commission to the European Parliament, the Council, the European Economic and Social Committee and the Committee of the Regions, Public Procurement for a Better Environment.

Construction Products Association (2012). Guidance Note on the Construction Products Regulation. (Available at: http://www.constructionproducts.org.uk. Accessed 29 June 2015).

Department of Business, Innovation and Skills (2014). Growth Dashboard. (Available at: https://www.gov.uk/government/uploads/system/uploads/attachment_data/file/396740/bis-15-4-growth-dashboard.pdf. Accessed 29 June 2015).

IDeA (2003). Sustainability and Local Government Procurement Improvement and Development Agency (IDeA).

Islam, M. and Siwar, C. (2013). A Comparative Study of Public Sector Sustainable Procurement Practices, Opportunities and Barriers. International Review of Business Research Papers, 9(3), 62–84.

Kenley, R., London, K. and Watson, J. (2000). Strategic Procurement in the Construction Industry: Mechanisms for Public Sector Clients to Encourage Improved Performance in Australia. Journal of Construction Procurement, 6(1), 4–19.

Love, P.E.D., Skitmore, M. and Earl, G. (1998). Selecting a Suitable Procurement Method for a Building Project. Construction Management and Economics, 16(2), 221–233.

McMurray, A.J., Islam, M.M., Siwar, C. and Fien, J. (2014). Sustainable Procurement in Malaysian Organizations: Practices, Barriers and Opportunities. Journal of Purchasing and Supply Management, 20(3), 195–207.

Mohan, V. (2010). Public Procurement for Sustainable Development, 4th International Public Procurement Conference, 26th–28th August 2010, Seoul, South Korea.

National Audit Office (2005a). Improving Public Services through Better Construction. The Stationery Office, London.

National Audit Office (2005b). Sustainable Procurement in Central Government. National Audit Office, London.

Newcombe, R. (2000). An Investigation into Simulating the Procurement Process in the United Kingdom Construction Industry. Journal of Construction Procurement, 6(2), 104–120.

Ochoa, A., Fuhr, V. and Gunther, D. (2003). Green purchasing in practice: Experiences and new approaches from the pioneer. In: C. Erdmenger, ed. Buying Into the Environment. Greenleaf Publishing, Sheffield.

Peiris, V.R.S. (2013). Sustainable Supply Chains in Sri Lanka, 3rd Global Networking Conference on Resource Efficient and Cleaner Production in Developing and Transition Countries, 4th–5th September 2013, Montreux, Switzerland.

Rowlinson, S., Matthews, J., Phua, F., McDermott, P. and Chapman, T. (2000). Emerging Issues in Procurement Systems. Journal of Construction Procurement, 6(2), 90–103.

The Joint Contracts Tribunal (2014). Procurement. (Available at: http://www.jctltd.co.uk/procurement.aspx. Accessed 29 June 2015).

Theron, C. and Dowden, M. (2014). Strategic Sustainable Procurement: Law and Best Practice for the Public and Private Sectors. Dō Sustainability, UK.

The Sustainable Procurement Task Force (2006). Procuring the Future – Sustainable Procurement National Action Plan: Recommendations from the Sustainable Procurement Task Force. DEFRA.

Thomson, J. and Jackson, T. (2007). Sustainable Procurement in Practice: Lessons from Local Government. Journal of Environmental Planning and Management, 50(3), 421–444.

United Nations (2006). UN Procurement Practitioner's Handbook. (Available at: https://www.ungm.org/Areas/Public/pph/ch04s05.html. Accessed 29 May 2015).

Walker, H., Mayo, J., Brammer, S., Touboulic, A. and Lynchohan, J. (2014). Sustainable Procurement: An International Policy Analysis of 30 OECD Countries. 5th International Public Procurement Conference, 14th–16th August 2014, Dublin, UK.

Sustainable construction

Adetunji, I., Price, A., Fleming, P. and Kemp, P. (2003). Sustainability and the UK Construction Industry – A Review. Proceedings of the Institution of Civil Engineers, Engineering Sustainability, 156(4), pp. 185–199.

Ashley, R., Blackwood, D., Butler, D., Davies, J., Jowitt, P. and Smith, H. (2003). Sustainable Decision Making for the UK Water Industry. Proceedings of the Institution of Civil Engineers, Engineering Sustainability, 156(ES1), pp. 41–49.

Casella Stanger, DTI, Forum for the Future and Carillion (2002). Sustainability Accounting in the Construction Industry. CIRIA Publishing Services, London.

Cheung, S., Lam, T., Wan, Y. and Lam, K. (2001). Improving objectivity in procurement selection. Journal of Management in Engineering, 17(3), 132–139.

CIB (1999). Agenda 21 on Sustainable Construction, 237. CIB.

CIRIA (2008). CIRIA – The Construction Industry Research and Information Association. (Available at: http://www.ciria.org.uk. Accessed 29 June 2015).

Constructing Excellence (2004a) Sustainability (Available at: http://www.constructingexcellence.org.uk. Accessed 29 June 2015).

Constructing Excellence (2004b). Respect for People Equality and Diversity in the Workplace. (Available at: http://constructingexcellence.org.uk/resources/respect-for-people-equality-and-diversity-in-the-workplace-toolkit/. Accessed 29 June 2015).

DETR (2000). Building a Better Quality of Life a Strategy for More Sustainable Construction. DETR, London.

Egan, J. (1998). Rethinking Construction: Report of the Construction Task Force. DETR, London.

Gordon, M. (1996). Structuring Sustainable Construction Contracts. First Transportation Specialty Conference, Edmonton, Alberta, Canada, pp. 91–103.

Harvey, R.C. and Ashworth, A. (1997). The Construction Industry of Great Britain, 2nd edn. Read Educational and Professional Publishing Company, Oxford.

Hill, R.C. and Bowen, P.A. (1997). Sustainable Construction: Principles and a Framework for Attainment. Construction Management and Economics, 15(3), 223–239.

Hillebranbdt, P.M. (2000). Economic Theory and the Construction Industry, 3rd edn. Macmillan Press Ltd, London.

HM Government (2005). Securing the Future – Delivering UK Sustainable Development Strategy. The Stationery Office, UK.

HM Government and Strategic Forum for Construction (2008). Strategy for Sustainable Construction.

Langford, D.A., Zhang, X.Q., Macleod, I. and Dimitrijevic, B. (1999). Design and Managing for Sustainable Building in the UK. In: S. O. Ogunlana, ed. Profitable Partnering in Construction Procurement: CIB W92 (Procurement Systems) CIB TG 23 (Culture in Construction) Joint Symposium. Chiang Mai, Thailand, January 1999 E & FN Spon, pp. 373–382.

Latham, M. (1994). Constructing the Team, HSMO, London.

Liu, A.M.M., Lau W.S.W. and Fellows, R. (2012). The Contributions of Environmental Management Systems towards Project Outcome: Case Studies in Hong Kong. Architectural Engineering and Design Management, 8(3), 160–169.

Ofori, G. (1998). Sustainable Construction: Principles and a Framework for Attainment – Comment. Construction Management and Economics, 16(2), 141–145.

Parkin, S., Sommer, F. and Uren, S. (2003). Sustainable Development: Understanding the Concept and Practical Challenge. Proceedings of the Institution of Civil Engineers, Engineering Sustainability, 156(1), 19–26.

Raynsford, N. (2000). Sustainable Construction: The Government's Role. Proceedings of the Institution of Civil Engineers, Civil Engineering, 138(S2), pp. 16–22.

Rethinking Construction (2003). Demonstrations of Sustainability. Rethinking Construction, London.

Rethinking Construction's Respect for People Working Group (2002). Respect for People a Framework for Action. Rethinking Construction Ltd, London.

Rhodes, C. (2013). Construction Industry: Statistics and Policy (Standard Note: SN/EP/1432). House of Commons, UK.

Sourani, A. and Sohail, M. (2011). Barriers to Addressing Sustainable Construction in Public Procurement Strategies. Proceedings of the Institution of Civil Engineers, Engineering Sustainability, 164(ES4), pp. 229–237.

Strategic Forum for Construction (2002). Accelerating Change – A Report by the Strategic Forum for Construction Chaired by Sir John Egan. Rethinking Construction, London.

Sustainable Development Research Network (2002). A New Agenda for UK Sustainable Development Research. Policy Studies Institute, London.

TCPA and WWF (2003). Building Sustainably: How to Plan and Construct New Housing for the 21st Century. WWF.

United Nations (2002). The Report on World Summit on Sustainable Development, Johannesburg, South Africa, A/CONF.199/20.

Uttam, K. (2014). Seeking Sustainability in the Construction Sector: Opportunities within Impact Assessment and Sustainable Public Procurement, PhD Thesis, Royal Institute of Technology, Sweden.

WCED (1987). Our Common Future. Oxford University Press, Oxford.

Xiao, H. and Proverbs, D. (2003). Factors Influencing Contractor Performance: An International Investigation. Engineering, Construction and Architectural Management, 10(5), 322–332.

3 Cost Modelling for Sustainability

Bernd Kochendoerfer, Margarete Roigk, Basak Guçyeter, H. Murat Gunaydin and Tofigh Tabesh

3.1 Introduction

Sustainability is a term that encompasses a broad definition, focusing on maintaining and improving the social, economic and environmental state for present and future generations (Brundtland, 1987; Ortiz et al., 2009). Sustainability involves methods that do not completely use up or destroy natural resources ("Sustainable" Def. 2. *Merriam Webster* Online). Since the 1970s, the methods to provide a sustainable built environment have been extensively discussed and explored among professionals and academics. The major and mutual emphasis of these methods usually draws attention to the excessive contribution of the built environment on energy consumption and relative adverse impacts on the environment. This mutual emphasis is often reciprocated with a reflex to frame the limits of sustainable constructions with realisation of sub-targets such as reduction of energy consumption during the operational life of a building, development of low energy designs and so forth. Such a response could be expected since sustainability is a contested and multi-dimensional context in search of a consensus in equally encompassing social, economic and environmental dimensions (Guy and Farmer, 2001; Bebbington et al., 2007).

The broad scope of sustainability is usually rendered to quantifiable aspects that are directly related to reduction of the energy consumed. Advances in a particular effort on energy conservation are a *cause célèbre* both in the context of an array of certifications/roadmaps and in technological developments. However, it turns out that each party involved in the sustainable built environment procedure must acknowledge a most beneficial aspect related to their position in the sustainable built environment. For instance, investors and owners in the construction industry could be concerned with attaining a sustainability goal directly in relation with the economics of construction, while for inhabitants or non-governmental organisations indoor environmental quality and consumption conservation could be preferential (Ding, 2008; Rodrigues and Freire, 2014).

In this regard, it is possible to state that extremely biased discussions on sustainability, the highly theoretical nature of these discussions and a lack of consensus on how to achieve a sustainable built environment, compel researchers towards quantifiable methodologies in assessing energy conservation measures, environmental impact assessment and financial characteristics of sustainability for the built environment (Frame and Cavanagh, 2009; Ortiz et al., 2009). The dispersed nature of sustainability efforts has been widely criticised for the last decade and the research on the sustainable built environment now starts to abandon its needless reflex to frame sustainability that originates from simple benefit oriented viewpoints and the normative ideal to attain the ultimate goal. Instead, it is becoming widely recognised that, to achieve a sustainable built environment, a common ground for consensus is a better medium, which can represent all related parties and facilitate them with the opportunity to balance their own benefits with the other involved parties. The basic goals of sustainability are emerging and becoming clear through stakeholder interaction and public awareness.

Achieving such a multi-dimensional approach within sustainability guidelines for built environment requires not only energy conservation goals, but also appropriate cost methodologies in order to ensure economic sustainability as well. Hence, the approach would be able to facilitate interaction between different parties involved in the sustainable built environment. Most efforts may therefore find the required consensus for sustainable design strategies, through being able to simultaneously minimise the cost and environmental impact of a building (Antipova et al., 2014). Systems thinking may also help to understand the interdependencies between the guidelines and methodologies.

The main determinant, however, is the cost of construction work. Any sustainability effort in the construction industry is evaluated through costs. In today's world, it is certain there is an absolute necessity both to sustain environmental resources and to provide cost efficient applications. Methodologies, such as Life-Cycle Analysis (LCA), introduce multidisciplinary aspects as full-cost accounting becomes favourable over time, since there is the capacity of balancing inputs and outputs through quantitative data (Ortiz et al., 2009).

Holistic life-cycle assessment methods that integrate cost, energy and environmental assessments should be the basis for sustainable decision making for the built environment (Loftness et al., 2003; Watson and Jones, 2005). In addition, as Verbeeck and Hens (2010) emphasise, life-cycle calculation models developed as a holistic methodology to optimise buildings should simultaneously consider thermal comfort and indoor air quality as fundamental inputs for more effective sustainable solutions for the built environment.

The ethical aspects underline the necessity for much more balanced economic and innovative solutions. One might consider the impacts of sustainability methodologies on the individual, group and business levels. The sole purpose here at the ethical level is to optimise benefits of sustainability efforts for all stakeholders involved as well as all living organisms and the earth. Systemic thinking and mutual understanding via productive communication is *sine-qua-non* for the continuous and sound development of our civilisation.

In order to elucidate essential factors of these discussions, this chapter focuses on the exploration, determination and optimisation of life-cycle costs for sustainable construction costs and presents their relationship with energy conservation and management. This chapter covers definitions such as cost, cost accounting, accounting for sustainability and introduces methods and tools on how cost control systems are integrated in life-cycle considerations for the built environment. Evaluation of life-cycle costs as a decision parameter for holistic assessment of sustainability of the built environment is explained. In addition, exploration of research on life-cycle costs and their relationship with the sustainable built environment is provided as well.

3.2 Tools and methods of cost modelling for sustainability

3.2.1 Cost modelling for building constructions

In order to explain the methods of cost modelling for sustainability explicitly, it is necessary to define the constituent concepts such as cost, cost accounting and full-cost accounting. Given the scope and definitions of these concepts, their relationships with the sustainable built environment can be clearly established.

Laitinen (2014) cites Horngren et al. (1999) to define cost as "a resource sacrificed to achieve a specific objective, usually expressed in monetary terms". Costs are determined by multiplying a quantity factor with a cost parameter. Depending on the reference unit, a differentiation can be made between planning-oriented and implementation-oriented methods. In the context of constructions, cost refers to all arrays of overall building design, including materials, investments, workmanship, amount of energy consumed during the construction of a building, maintenance, operation and so forth (prEN15459, 2006).

Cost accounting, one of the two major fields of accounting discipline, is a cost information system, which measures and financially records all costs and associates them to a product as an economic value (Nigam and Jain, 2001). It is one of the main methods in developing financial management decisions such as product pricing based on production costs for different products and the attributed selling price of these products (Jasch, 2003). Despite the fact that cost accounting is a widely accepted traditional cost model, it lacks inclusion of demolition and recycling costs of a product in the economic valuation process (Gluch and Baumann, 2004). Cost accounting, therefore, is capable of carrying out the classic accounting methods and helps designers and investors to assess nominal values for production and operation of the whole building, yet it fails to integrate externalities directly related to the life of a building in a holistic manner.

Full-cost accounting on the other hand, is a methodology that includes the end of service life costs and recycling costs to the overall service life costs of a product and may be considered as the externalities for cost accounting. With its characteristics that move beyond the traditional accounting methods and includes benefit analysis and externalities valuation, full-cost accounting has become a methodology engaged with sustainable development principles (Bebbington et al., 2007; Bebbington and Larrinaga, 2014).

The previously mentioned accounting concepts are directly related to the economics of building constructions. For instance, even in the basic sense, the affordability of a building design for an owner or an investor can be evaluated through these methods. For a specific design project it is possible to break down overall costs to lower working levels and certain time spans in which the costs are incurred and the cost type is classified. Here it is possible to differentiate between auxiliary, material or overhead costs. Since the cost elements are determined, parameters are assigned to quantify elements. Briefly, two methods; engineering and parametrics are explained, respectively, and more in detail.

The engineering method is based on a direct estimate of the individual cost elements. Here, only costs for development and manufacture are taken into account. For instance, if the material costs for a concrete wall have to be determined, using the volume and an estimate of the planner of the unit price, the cost of the element is calculated. For this method, the experience gained during similar projects is used as a reference. Taking into account technological progress, inflation, price changes and other factors, cost elements are adapted from previous projects. This method is characterised by a low level of complexity and low costs, as the procurement of data is straightforward (DIN EN 60300, 2005).

The parametric model determines parameters and variables that affect the cost elements and determine their magnitude. The building project-specific factors primarily affect the builders, while the costs are based on the pricing policy evaluation of the market situation, depending on economic, construction and location related parameters. The amount of the costs is also influenced by user-specific factors (DIN EN 60300, 2005).

The focus, however, is on object-specific factors. These factors include the building type, geometry, construction methods, equipment used and choice of materials, government regulations and statutory requirements. All these features, grouped under object-specific parameters, have a strong effect on the costs during the construction and service life of a project.

Given a basic idea for cost modelling for building construction, the latter sections aim to expand on the subject in the light of cost accounting concepts. Cost modelling is discussed beyond static methodologies and the evaluation of how to include constraints such as time, energy performance, environmental impacts offers an in depth exploration of these defined concepts.

3.2.2 Life-cycle costs for building construction

The amount of energy consumed through the service life of buildings is directly related to the technologies utilised in construction and the type and amount of building materials used. In general, the high cost building materials (e.g. insulation materials) result in lower energy consumption costs during the operational period of the building, as a result of lower energy losses (Pajchrowski et al., 2014). In addition, Verbeeck and Hens (2010) argue that the efforts in improving the energy performance of buildings results in extra use of materials and components, thus the embodied energy, energy needed for production and transport of all these materials and components, increases. However, in today's world, to reduce the acceleration of environmental problems such as global warming, ozone depletion and so on, there is an urgent necessity to act upon appropriate methods in sustainably balancing both construction costs and energy consumed by the built environment.

Building energy efficiency (or performance) background has long been studied notably in practice and theory. Hence, with the trends in research, it is possible to observe the necessity to evaluate and integrate cost of buildings through their life cycle as well. In addition, the effects of cost methodologies on achieving the desired sustainable built environment is worthy of further evaluation due to their potential in conserving financial resources. In order to conduct such an evaluation it is necessary to understand the concept of cost in the life cycle of buildings and how it could be integrated into holistic sustainability approaches.

The term "life cycle", when used for buildings or construction projects, refers to the entire life course of a building. Life-cycle methods do not only focus on the operational period of a building, but also take design, construction, operation, demolition and waste treatment phases into consideration. Life-cycle assessment

(LCA) is a tool that facilitates systematic analysis of environmental performance of products or processes over their whole life cycle (Cabeza et al., 2014). In other words, LCA is an environmental management methodology that evaluates the environmental impacts of a product or service, starting from extraction of raw materials, manufacturing, production and use and finishing with final disposal; that is, from "cradle to grave" (Ortiz et al., 2009).

LCA is a widely accepted and utilised tool for assessing the environmental impacts of the building industry. LCA is commonly used in the classifications listed next. These classifications are reviewed from the literature by Cabeza et al. (2014), Ortiz et al. (2009) and Singh et al. (2011):

- Construction product selection.
- Construction systems/process evaluation.
- Tools and databases related to the construction industry.
- Methodological developments related to the construction industry.

ISO 14040, Environmental management – LCA – Principles and framework (2006), is an environmental management standard adopted by The International Organization for Standardization (ISO), which focuses on establishing methodologies for LCA. The standard includes definitions for steps of LCA, which provides a better understanding on the extensive scope of the methodology. The standard defines and differentiates the first two steps as LCA and LCI (life-cycle inventory analysis). The former evaluates inputs, outputs and the potential environmental impacts of a product system throughout its life cycle, while the latter is "a phase of life cycle assessment involving the compilation and quantification of inputs and outputs for a product throughout its life cycle". The third and fourth steps are life-cycle impact assessment (LCIA) and life-cycle interpretation and are, respectively, defined as; (3) "phase of life cycle assessment aimed at understanding and evaluating the magnitude and significance of the potential environmental impacts for a product system throughout the life cycle of the product" and (4) "phase of life cycle assessment in which the findings of either the inventory analysis or the impact assessment, or both, are evaluated in relation to the defined goal and scope in order to reach conclusions and recommendations" (ISO14040, 2006).

Not just the environmental impacts could be evaluated through LCA methodology, but also the systematic approach of the methodology is utilised to develop models such as life-cycle energy analysis (LCAE) and life-cycle cost (LCC) analysis (Cabeza et al., 2014).

As this chapter focuses on the cost modelling methods for sustainability, it is necessary to define life-cycle cost analysis. Fuller (2014) defines life-cycle cost (LCC) analysis as a method to assess total cost of facility ownership and adds that the analysis "takes into account all costs of acquiring, owning and disposing of a building or building system".

LCC analysis could be utilised to compare, evaluate and select the optimal alternative among a set of alternatives that accomplish same performance requirements, hence it differs regarding their initial costs and operating costs. In such situation LCC could be used to compare and contrast project alternatives and allows a decision model in order to select the one that maximises net savings. LCC might as well be utilised to compare different project drafts to estimate profitability of projects and identification of cost drivers.

Evaluation of life-cycle costs within any holistic sustainable built environment efforts is crucial as the means to elucidate the optimisation of financial aspects and contribute to the economics of environmental sustainability. Earlier, conventional planning and construction methods solely focused on the construction costs of a building as simple investment – revenue balances. Despite the fact that the opportunities for optimisation and decision making are greatest in the project development and planning phases, subsequent costs were given little attention (DGNB, 2009). Therefore, the objective of holistic planning emerges as a vital contribution for optimisation of the life cycle of a building by taking both economic and energy based measures into account. Including life-cycle costs in assessment models would facilitate optimisation for energy, emissions and cost simultaneously.

The calculation of life-cycle costs is based on related costs compounded over numerous years. The life-cycle costs here represent the sum of all expenditures in individual years. Due to the long lifespan of real estate objects, the time at which the payments are made is significant for the calculation results and must be taken into account accordingly (Pelzeter, 2006).

To account for the time-dependent value of the incoming and outgoing payments, they are either discounted or compounded to the valuation date. Using profitability accounting[1] it is possible to determine whether a measure pays for itself, for example, could higher investment sums in the construction phase effect energy savings and thus lead to lower operating costs during the operational phase? (Pelzeter, 2006). The magnitude of the life-cycle costs also depends on factors such as location, type of building use, number of unit, spatial design, structural design, construction standard and flexibility of use (Möller, 2001).

The period under consideration for a building depends greatly on the type of building and its function. In particular, the quality and durability of the selected building materials can lead to a longer lifespan and a reduction in maintenance costs (BMVBS, 2013). Decisions that are relevant for costs of subsequent use, renovation and revitalisation measures and disposal of the building are made in the planning and realisation phase. Changes made later on can only be realised at considerable cost and effort (Pelzeter, 2006).

3.2.3 Determination of life-cycle costs

In order to determine costs of a product or process within its life cycle, it is necessary to identify a framework to cover the cost breakdown structure (CBS) of that specific product or process. CBS is a set of categories in which costs constitute categories based on different organisational levels (Fabrycky and Blanchard, 1991). Different guidelines for life-cycle cost accounting use different phases to obtain the breakdown structure. For instance, GEFMA (German Facility Management Association) guideline 200 "Costs in facility management" proposes a CBS that spans the entire life cycle as follows; conception, planning, construction, marketing, procurement, operation and use, renovation and restoration, vacancy and recovery (GEFMA, 2005).

The ISO standard, EN 60300-3-3 (European Norm) on the other hand, defines six phases for CBS for a product's life cycle. The first phase covers conception and definition, where ideas are collected for a new project and initial studies are prepared. The second phase, outline and development, includes preparation of financing concepts, construction plans and schedules. After planning and obtaining relevant approvals, a request for tender is completed and is followed by the following two steps: (3) assembly and (4) installation. The fifth phase is crucial for the entire project, as the greatest costs generated in the shortest periods are influential on the longest phase of the product, the operation and maintenance phase. Disposal is the last phase of a life cycle, in which concepts are developed and realised regarding how disposal will be performed (DIN EN 60300, 2005; Hoffart and Hirsch, 2011).

Fabrycky and Blanchard (1991) classify life-cycle costs in a breakdown structure based on organisational activities, proposing four main categories. The first category, *research and development costs*, includes activities such as initial planning, feasibility studies, product research, analysis of requirements and so on. (Fabrycky and Blanchard, 1991). This category refers to the pre-evaluation and preparation phase when the main decisions are made based on the requirements of the project. Even though this category might be perceived as a preparatory phase, it has a direct and permanent influence on future organisational activities that relate to these preliminary decisions. The second category, *production and construction costs*, consists of organisational activities such as manufacturing (fabrication and assembly), facility construction, production processes, quality control, logistic support and so on (Fabrycky and Blanchard, 1991). The activities in this category relate to the production and assembly of the project components. It covers all the costs that occur during the production phase of a project, including sub-activities such as transportation and management during construction. The third category, *operation and maintenance costs*, includes user operations, maintenance of the product, modifications, facilities and so on (Fabrycky and Blanchard, 1991). The activities in this category relate to the operational period of the specific product or system. It covers all the costs that occur during the use of the system or product; that is, the costs of maintaining and running it. The fourth and final category, *retirement and disposal costs*, includes recycling, disposal and any necessary logistic support requirements (Fabrycky and Blanchard, 1991). The costs in this category are the result of the end-of-service-life procedures. This category covers all the costs related to the demolition, sorting, recycling and disposal of the product and also includes the costs of sub-activities such as the transportation of material and management of the disposal phase.

Although the phases or organisational activities summarised previously do not refer to a specific context, all the phases (and related costs) could be adapted to the construction industry.

In determination of life-cycle costs, time of origin for assessment has a significant impact on the future recurring costs (e.g. operational costs). An earlier assessment in the life cycle of a product of the cost factors has a greater impact on the overall life-cycle cost estimations. Thus, odds of obtaining an increasing profit are optimised. Practical experience has shown that at the end of the concept and definition phase, half the relevant costs of a product have already been determined (DIN EN 60300, 2005).

In architectural design and engineering, the opportunity to influence life-cycle cost estimations is higher in the basic investigation phase. From this period on, influence of attempts to increase profits for life-cycle costs of a construction decreases incrementally (Pfarr, 1976). Even at this early phase of the life-cycle cost assessment for building constructions, a precise formulation of the factors is necessary to determine a benefit-optimised solution. A systematic approach is necessary to achieve a benefit-optimised decision based on the life-cycle costs.

The foundation for every analysis is the definition of the objective. In cases that determination of life-cycle costs is employed as a model for profitability assessment, objectives are formulated to achieve an optimised solution. After the formulation of objectives, data collection is the next step, in which data should be collected for every cost element and should be allocated according to product alternatives and subsystems. Based on the data, life-cycle analysis is performed by taking into consideration the prerequisites from the analysis plan. The resultant data are classified and summarised in logical and user-friendly categories such as fixed and variable costs, one-time and recurring costs, and procurement, operating and disposal costs.

With the help of a sensitivity analysis, the impact of all the assumed parameters and decisions could be evaluated. In this context, the input and output variables of the performed analysis are scrutinised. Finally, an assessment is made whether the pre-defined objectives were achieved and sufficient information has been obtained to make a final decision (DIN EN 60300, 2005).

3.2.4 Models for cost determination

Costs are determined by multiplying a quantity factor with a cost parameter. Depending on the reference unit, differentiation can be made between planning-oriented and implementation-oriented methods.

Planning-oriented methods determine the costs using function related (e.g. building programme), building related (e.g. gross floor area) or construction element-related properties (e.g. cost groups according to DIN 276-1 (2008)).

Implementation-oriented methods calculate costs as they occur during implementation. These could relate to partial services or contract units. For the implementation-oriented method, all costs can be determined for the provision of services, for example using a calculation of quantity and value. Here, the partial services, direct individual costs, overhead costs of the construction site, general business expenses and profit and risk are considered. In the construction industry, the calculation using the total (apportionment procedure), the calculation with pre-determined surcharges (surcharge calculation) and the calculation with hourly rates are used.

For investment analysis, typically with the help of methods, the objective function of the decision problem is developed using certainty and then the reliability of the outcome or the outcome distribution is determined using the methods under uncertainty. Jacob (1994) classifies the methods for investment calculation for individual decisions into two main categories: (1) models with certainty and (2) models with uncertainty.

Models with certainty

Models with certainty could be divided into statistical and dynamic methods. Statistical methods relate exclusively to a period, for example a certain period of use or an average period. The temporal difference in incurring costs and the different value of the payments are not taken into account in these methods. Examples for statistical methods are the cost comparison, the profit comparison, the profitability comparison and the statistical payback calculation (Jacob, 1994). Since temporal factors are ignored, these methods are less suitable for life-cycle cost calculation and thus are not widely used.

Dynamic methods consider the object-specific incoming and outgoing payments directly and not per period. Thus, multiple period analyses are possible, making the various payments comparable. Dynamic methods can be divided into asset value and interest rate methods (Jacob, 1994). While the interest rate method determines the rate at which the fixed capital bears interest, the asset-based methods make it possible to evaluate the wealth generated of various investment alternatives (DIN EN 60300, 2005).

Models with uncertainty

Due to the unpredictability of the future, data of life-cycle costs analyses might contain uncertainty. Uncertainty can occur in the form of the unknown or risk; for risk the probability of occurrence exists, which is not the case for the unknown. For assessment of building life-cycle costs, sensitivity analysis and a risk analysis are the most commonly used methods (Jacob, 1994).

For sensitivity analysis, the uncertainty of the input variables varies to evaluate the influence of the changes on the target variable, while a risk analysis attempts to determine the probability distribution of the target variable. Stochastic[2] dependencies between uncertain output variables and the target variable are observed with the help of correlation coefficients.

After the cost element and influence parameters have been identified, the evaluation of the alternative courses of action is performed. Separate collection of material, wage, equipment and disposal costs makes it possible to select between and recombine the various planning and implementation variations of individual construction phases to achieve the full optimisation potential. The final decision is made by the principal or by an authorised third party (DIN EN 60300, 2005).

3.3 Interdisciplinary methods for optimisation of life-cycle costs

In the Report of the United Nations Conference on Environment and Development (UN, 1992) Principle 1 states that: "Human beings are at the centre of concerns for sustainable development. They are entitled to a healthy and productive life in harmony with nature." Since then, nation states, particularly industrialised countries, have sought to develop guidelines and create a framework for anchoring and promoting sustainable development.

With the manifestation of the concept of "sustainability", it became more vital for the public to reflect on the optimum utilisation of finite resources and the means for bringing out a distinctive attitude to protect those resources. The assets of the society, such as environment, resources, health, culture and capital, must be protected.

Construction, renovation and operation processes of buildings should be economically, ecologically and socially sustainable (BMVBS, 2013) in order to achieve the goals for a sustainable built environment. The most important consideration in attaining these goals is to build up a holistic approach. The building life cycle, which can be roughly divided into phases of product manufacture, installation, utilisation and dismantling, is thus thrust into the centre of sustainable development. Depending on the type of building and the type of use, considerations of long-term cost savings across all phases are necessary at an early stage. Sustainable building, therefore, means the minimisation of all energy and resource use as well as minimal impact on the ecosystem across all life cycles.

Since the 1980s, the development and use of new technologies and construction methods have been promoted and implemented with the purpose of obtaining a sustainable built environment. Familiar terms such as low-energy house, passive house, zero-energy home, plus energy home or energy efficient building have emerged since then and were expected to have significant advances on the development of sustainable buildings (Ebert, 2010). However, these terms are based on concepts such as efficiency and zero-energy; there is still necessity to cover a uniform approach and a framework that integrates the complex interaction of a number of factors, such as different interest groups, use of current technologies, cost distribution and assumption of costs, funding guidelines and programmes, type and duration of use, holistic life cycle and evaluation of the future and suitable considerations. With an evaluation system or certification system, the various aspects and manifestations of sustainable construction should be made measurable, comparable and transparent.

Since LCA models are managed through the breakdown of organisational phases and each phase is defined through inventory sets of sub-elements, these models could be controlled to achieve the capability to be integrated within interdisciplinary approaches, as holistic assessment models. These holistic assessment models can be perceived as a sum of simultaneously functioning assessment models, which can be related to environmental impacts, primary energy consumption, management tools and so on. The following subsections summarise the life-cycle approaches that are independent in terms of their associated outcomes, yet have the potential to become a part of a holistic process. It is also important to bear in mind that the following concepts of life-cycle evaluations have the potential to correspond with a cost analysis, which has to be taken into account during the determination of life-cycle costs in a holistic approach.

3.3.1 Life-cycle engineering

Life-cycle engineering (LCE) emerged in response to the exploration of holistic methodologies based on assessment of product life cycles with least possible environmental impacts and financial feasibility simultaneously (Ribeiro et al., 2008). LCE does not only employ traditional approaches for performance evaluation, but also includes life-cycle tools to analyse costs and environmental impacts for assessment of sustainability measures (Finnveden and Moberg, 2005).

LCE is especially important in integrating cost considerations, since resource consumption by utilities – such as electricity, energy, water and waste management – represent a significant proportion of operating costs. When one examines the way the costs are accrued in the individual phases, depending on the duration and type of use, it can be observed that the costs incurred during the operational phase can exceed the costs accrued during the construction phase by a large factor. The 80:20 rule frequently refers to this. This means that only 20% of the costs are incurred during planning and manufacture, while 80% of the costs arise during the operational phase. The costs for demolition and recovery are neglected in this analysis (Hellerforth, 2006). The cost drivers are primarily the costs for operation and administration, for maintenance and for partial dismantling and renovation procedures that result from the changing requirements of the users.

3.3.2 Facility management

Facility management can be considered as a part of the life cycle of a building, in which the sum of all operations during the occupancy phases of a building is managed or accounted for. Facility management includes control and maintenance of the services that are critical for the operation of building facilities, such as building fabric, ventilation systems, electrical installations (Lai and Yik, 2011) and so on. These building services are directly influential on the consumption patterns of the building, occupant comfort and environmental impacts. Therefore, their management and control is relatively influential on the recurring costs and impacts of a building.

All approaches and definitions for facility management have the concept of the life-cycle approach in common. Thus, user requirements should be identified and guided at an early stage to support all value-adding processes during the operational phase efficiently and cost effectively. The associated services can be controlled and monitored centrally from a single location. Facility management is divided into the four areas; infrastructure, commercial, technical management and space management. The objective of facility management is thus adapting the use of resources optimally pertaining to the needs of the user and managing the real estate object as feasibly as possible (Braun, 2004).

The characteristics of facility management demonstrate the importance of the design of all life stages of a building object above all. In addition, facility management typically has the knowledge from corresponding reference projects and offers owners or users of an object a good alternative for gaining insight into sustainability and possible savings potential at an early stage.

Facility management provides a holistic view of people, technology, services and property. By taking into account political, economic and social factors, these four aspects should be co-ordinated and harmonised. The level of transparency created via the processes, consumption and costs during the operational phase promotes fast and targeted intervention into the existing system (Nävy, 2006). It also contributes to flexibility with respect to the usability of an object. Due to changing user requirements or environmental conditions in time, expensive and complicated changes to the building structure could be anticipated.

Facility management organises the direct provision of the required infrastructure so that a change in utilisation, exact accountability and controlled removal could be made possible. With appropriate selection of construction products and suitable use, the basis for optimisation of the maintenance and repair costs could be laid at an early stage. With the implementation of these measures, a permanent reduction of the operational costs is possible. Thus, the economic benefits for the user can be highlighted and placed in the foreground. Moreover, it will also be possible to influence the technical lifespan of an object (Kyrein, 2002).

Previously, the focus was more on reducing the primary energy demand through the utilisation of modern technologies. At the same time, however, importance should be attached to the ability to convert the usage of the building. Another important aspect is the suitable use of construction products, corresponding to their strengths. Long-term and sustainable functionality of the components should thus be secured, both individually and in combination.

This evaluation reveals that facility management represents a significant contribution and an important instrument to address the ideas of sustainable building and to shape it actively. Through the holistic approach and core orientation of facility management, for instance, all topics of the DGNB (2009) certification system are addressed. Depending on the type of operation, the available budget and the need for a sustainable point of view would be taken into account in different degrees. By involving facility management in the project programme at an early stage, a substantial contribution could be made to the fulfilment of individual criteria. Successful certification is not automatic, as the number of criteria goes beyond the scope of facility management.

3.3.3 Primary energy

The fundamental goal of climate policy is to limit the primary energy demand for non-renewable energy sources. Directive 2010/31/EU of the European Parliament and of the Council on the Energy Performance of Buildings (EPBD, 2010) defines primary energy as the "energy from renewable and non-renewable sources which has not undergone any conversion or transformation process". The term *primary energy* is used to indicate the sum of energy required to produce a product, including all energy resource inputs and losses along the energy chain (Gustavsson and Joelsson, 2010).

The Energy Performance of Buildings Directive (EPBD, 2010) declares that the energy performance of a building could be evaluated with an "energy performance indicator and a numeric indicator of primary energy use, based on primary energy factors per energy carrier, which may be based on national or regional annual weighted averages or a specific value for on-site production". EPBD points out the primary energy as a means of evaluation for consumption patterns in the operation life of the building. However, life-cycle analysis studies investigate primary energy consumption beyond the energy use for operational phase and integrate the primary energy utilisation for production of building materials, transportation and so on.

In order to promote efficiency of buildings, primary energy utilisation through all life-cycle steps should also be evaluated in a breakdown of organisational activities. This approach is vital since reduction of the final or purchased energy during the operational life of a building does not correspondingly reduce the natural resource consumption or the emissions during the life cycle of a building (Gustavsson and Joelsson, 2010).

The primary energy demand depends on whether non-regenerative energy sources are used or on the ratio of primary energy from non-renewable and renewable energy sources used. With respect to the non-renewable portion of primary energy, it must be determined how this primary energy is generated and how high the ultimate demand for final energy is. The final energy is composed of various quantities of energy such as the heating and water heating demand (also called useful energy), including losses in the system equipment itself. If you multiply the sum of all useful energy consumed with the system expenditure factor, the result will give the primary energy demand. The system expenditure factor includes the primary energy factor, which describes, in addition to the actual energy value of an energy source, the energy required for extraction, production and transport of the product. Which final energies will ultimately be needed and in what quantities depends in part on the demand and operation scenarios. The heating demand is crucial in any case because it represents the greatest proportion of energy demand in residential buildings and usually in non-residential buildings as well. The heating demand depends on the transmission heat loss HT [W/m²K]. The transmission heat loss is primarily dependent on material property, the heat transmission coefficient U [W/m²K] of the component facing outside. Thus, there are various elements which affect the primary energy demand of a building (ISO14040, 2006).

3.3.4 Scope of inventories and rating systems in LCA

There can be various motivations for preparing an LCA; hence, the objective should be to select the ecological friendly materials for the production of a building as long as they are compatible with technical and economic considerations. All other steps of LCA work towards this goal and are influenced by it.

The inventory is the compilation of quantifiable environmental factors (which were already limited in the scope), also called input and output factors. Input factors are all product, material and energy flows that must be supplied to an LCA unit so that it is created, while output factors are the flows that the unit gives off to the environment, such as emissions or waste.

Since, as described, we should be dealing with a dynamic rating system with reference values, the objective should not merely be to obtain a lesser value, but rather to undercut the fluctuating reference value of the primary energy demand of its counterpart, the actual building. The reference building always corresponds to the actual building with respect to volume, geographic location, orientation and use; only the system equipment and quality of the building envelope are fixed. For non-residential buildings, a differentiation is made between multiple zone models (standard procedure for office and administrative buildings) and individual zone models as, unlike a residential building, a non-residential building can be divided into different areas of use (office, meeting, servers etc.). Not only heating, but also water heating and ventilation influence the annual primary energy demand of non-residential buildings; cooling and lighting are also included in the calculation (additional final energy consumption).

The input variables for the primary energy demand of the construction result from the mass of the materials used multiplied by the corresponding factor for the use of primary energy per reference unit of the material. The basis for this approach is an upstream LCA according to DIN EN ISO 14040/DIN EN ISO 14044. "The life cycle assessment refers to environmental aspects and potential environmental impacts (…) during the life cycle of a product from the extraction of the raw material to production, use, waste treatment, recycling to final disposal (i.e. from 'cradle to grave')" (ISO14040, 2006).

The scope defines the level of detail of the comparison, which is the order of magnitude at which the objects are assessed as well as the factors that are included in the approach. The former could be individual materials; as they typically are associated with other materials in construction, forming a part. Here, components are typically compared with others that have the same functional unit; an example would be the exterior wall as thermal insulation system or aerated concrete. The latter means how many production processes of a material or a part are included in the assessment; for example, the energy expenditure for the extraction of raw materials for transport, the actual production of the building material or the eventual installation of the material. To a similar extent, this also applies to maintenance/repair and disposal or recycling of a material. In addition, criteria are selected with which materials or components are compared; these are, for example, impact categories such as acidification potential or the effect on the greenhouse effect as well as the consumption of non-renewable primary energy demand or the amount of waste produced, which indirectly describe the potential environmental damage.

3.3.5 Impact assessment

Impact assessment is a set of results for the evaluation of the life-cycle inventory with respect to their importance for potential environmental effects. In other words, the life-cycle impact assessment determines potential damage caused by the system as defined by the quantities of inputs and outputs. In the interpretation phase, the inventory and impact assessment results are reviewed and conclusions are reached. Finally, recommendations on the environmental performance of one product relative to an other can be made (Srinivasan et al., 2014).

European standard EN 15804:2012, which includes category rules for all construction products and services, states that impact assessment should employ evaluation of seven environmental impact categories: global warming, ozone depletion, acidification of soil and water, eutrophication, photochemical ozone creation and depletion of abiotic resources (EN 15804, 2012). Global warming potential and the proportion of non-renewable primary energy demand have significant effects on the overall assessment of the other impact categories.

United States Environmental Protection Agency defines life-cycle impact assessment (LCIA) as a phase of a life-cycle inventory, in which potential human health and environmental impacts of resource consumption are evaluated (EPA, 2014). In order to assess environmental impacts, key steps of a life-cycle impact assessment can be listed as follows:

1   Selection and Definition of Impact Categories: identification of environmental impact categories (e.g. global warming, acidification etc.).
2   Classification: assignment of life-cycle inventory results to impact categories (e.g. classifying carbon dioxide emissions to global warming).
3   Characterisation: modelling life-cycle inventory impacts within impact categories using science-based conversion factors (e.g. modelling potential impact of carbon dioxide and methane on global warming).
4   Normalisation: expressing potential impacts in ways that can be compared (e.g. comparing the global warming impact of carbon dioxide and methane for the two options).
5   Grouping: sorting or ranking the indicators (e.g. sorting the indicators by location: local, regional and global).
6   Weighting: emphasising the most important potential impacts.
7   Evaluating and Reporting Life-Cycle Impact Assessment Results: gaining a better understanding of the reliability of the life-cycle impact assessment results (EPA, 2014).

The United States Environmental Protection Agency defines impact categories in relation to their associated endpoints and impacts as follows:

- Global Impacts: Global warming (polar melt, soil moisture loss, longer seasons, forest loss/change and change in wind and ocean patterns), ozone depletion (increased ultraviolet radiation) and resource depletion (decreased resources for future generations).
- Regional Impacts: Photochemical smog ("smog", decreased visibility, eye irritation, respiratory tract and lung irritation and vegetation damage) and acidification (building corrosion, water body acidification, vegetation effects and soil effects).
- Local Impacts: Human health (increased morbidity and mortality), terrestrial toxicity (decreased production and biodiversity and decreased wildlife for hunting or viewing), aquatic toxicity (decreased aquatic plant and insect production and biodiversity and decreased commercial or recreational fishing), eutrophication (nutrients – phosphorous and nitrogen – enter water bodies, such as lakes, estuaries and slow-moving streams, causing excessive plant growth and oxygen depletion), land use (loss of terrestrial habitat for wildlife and decreased landfill space) and water use (loss of available water from groundwater and surface water sources) (EPA, 2014).

Impact categories explained here present how the characterisation factors should be translated from different inventory inputs to facilitate direct comparison between diverse impact indicators (EPA, 2014). Even more comprehensive cost studies might be considered coupled with impact studies for a sustainable built environment, focusing on the degree of environmental effects and their costs for future generations.

3.4 Current research on LCC analysis

In recent years, LCA research has been used extensively in quantifying the environmental impacts of the built environment. Research on LCA presents a broad array of applications, ranging from focusing on the production of a single construction material to urban projects, evaluating their environmental impacts, contribution to energy consumption and life-cycle costs.

The review of literature reveals that a considerable number of studies focus on the evaluation of the life-cycle analysis between environmental inputs (materials and energy) and outputs (waste and emissions) (Azari, 2014). On the other hand, there are also efforts to integrate cost considerations into life-cycle analysis studies in recent literature.

In the era of sustainable building and a tendency of introducing environmental factors into the decision-making processes, environmental costs related to the individual decision-making scenarios have become a more frequent additional criterion. Corgnati et al. (2013) mentioned that recently economic evaluations have made use of the life cost analysis as well as through the total net present value in a defined building life cycle. The main reference for the economic calculation methodology is the global cost calculation from the European Standard EN 15459 (2006). For instance, Nässén et al. (2012) compare buildings with concrete and wooden structural frames focusing on their lifetime carbon dioxide emissions as well as their total material, energy and carbon dioxide costs.

Research indicates that net present costs for the different buildings are also affected by the future energy supply system, even though the impact is small, especially compared to the total construction cost. The research concludes that it is unclear whether wood framed buildings will be a cost-effective carbon mitigation option and emphasises that further analyses of costs should be performed before prescriptive materials policies are enforced in the buildings sector.

Verbeeck and Hens (2010) conducted a holistic optimisation study on life-cycle energy, cost and environmental impact of buildings. They utilised a combined approach of generic algorithm and the Pareto concept in order to solve a multi variable problem with multiple objectives. They identified a hierarchy of approaches for minimising life-cycle energy, through improvement of insulation in the building, installation of an efficient heating system and integration of renewable energy driven heating (Karimpour et al., 2014).

Another study by Han et al. (2014) also puts an emphasis on the fact that capital cost and operating energy consumption are both major parts of life-cycle cost. The study suggests that annual energy consumption becomes more important when the building has a longer lifespan. In this case study, the operating energy cost takes the largest percentage of the life-cycle cost when the building has a lifespan more than 30 years. Optimal combination for annual energy cost highlights the material with best thermal properties, while cheap materials and smaller HVAC system size are emphasised in optimisation of capital cost. Therefore, a holistic life-cycle analysis model provides a better understanding of objectives for building system optimisation. It also suggests that combinations of optimised building components are dependent on the building lifespan. There are two energy-pricing scenarios considered in this study. The results indicate that the increasing prices will influence life-cycle cost and optimisation results. A case study of an office building shows that when the building lifespan is greater than 30 years, the cumulative annual energy consumption cost is projected to be higher than the initial construction cost. Finally, optimal component combinations vary with different lengths of a building's lifespan. For instance, wood window frames become the optimal component for less energy and maintenance cost when the building lifespan changes from 14 to 60 years.

Chel and Tiwari (2009) evaluate size and cost of photovoltaic power system components using an actual case study in New Delhi. Based on LCC analysis, capital cost and unit cost of electricity are determined for photovoltaic systems such as stand-alone and building-integrated photovoltaic systems. The effect of carbon credit on the economics of photovoltaic system indicates reduction in unit cost of electricity by 17–19% and 21–25% for stand-alone and integrated systems, respectively.

Pikas et al. (2014) compare office building fenestration design proposals with an approach to both energy efficiency and cost optimality. An open-plan generic single office floor is divided into five zones and simulated through a test reference year for Estonia. The main aim of the study is to determine the optimal external wall insulation thickness and assessment of cost optimal and most energy efficient solutions for the façade. The result pointed out that, for the cold Estonian climate, triple glazed argon filled windows with a small window to wall ratio and walls with 200 mm thick insulation are revealed to be energy efficient and cost optimal within 20 years.

Brás et al. (2014) evaluate benefits of using external thermal insulation composite systems and compare it with a double brick wall, which is a common traditional local construction technique. Their research compares building behaviour when constructed in three different climatic regions of Portugal. Two dwellings are examined as case studies in order to conduct a thermal study and an energy (electricity-based) cost analysis for different scenarios, such as (1) influence of the shading devices systems in the thermal losses and gains, (2) changes in the glazing quality and (3) influence of the replacement of openings with PVC by an extruded aluminium system.

It is possible to extend the current studies discussed in this chapter by the broad selection of the published research in recent decades. Instead, to be able to point out the major denominators in conducting life-cycle analysis and life-cycle cost studies, it would be helpful to summarise the literature review of Cabeza et al. (2014) in terms of lifetime of analysis, scope of the analysis, functional unit analysed, system boundaries and building typology. These major denominators may trigger further research or point out the gap in the broad research area of life-cycle studies.

Cabeza et al. (2014) conducted a review based on 62 research papers on case studies using LCA methodologies. The main classification for investigation of such a large research database covers four topics. First is LCA in the building industry including building material comparison, building decision support tools and assessment systems. Second is the LCA of buildings, which encompass buildings according to their functions, such as residential, non-residential and civil engineering structures. The third classification includes LCEA of buildings based on embodied energy and operating energy. The fourth and the final classification points out the main concept this chapter is based on, life-cycle cost assessment (Cabeza et al., 2014). It is remarkable to note that only one of the 62 studies employs LCC analysis as a methodology for obtaining a holistic perspective, which is the research by Nässén et al. (2012) that focuses on CO_2 emissions and costs. Approximately 40% of the reviewed research utilise LCEA of buildings and only two of them utilise life-cycle costs as decision criteria. This information is extremely significant for the future life-cycle studies where cost considerations should be integrated in the decision processes for achieving a sustainable built environment.

The range of reviewed research is also very extensive in terms of the functional unit that is considered as subject to LCC analysis. Seventy per cent of the research utilises a functional unit ranging from LCA of a single building element to whole-building assessment. The common approach is either focusing on the whole-building or a unit area of a building with one square metre of representative produced and operated area (Cabeza et al., 2014). For further research to include life-cycle cost assessment, whole-building approaches could be utilised to clearly demonstrate the relationship between financial, environmental and social aspects of the sustainable built environment.

As a final remark, it is important to point out that Cabeza et al. (2014) signify lifetime considerations of the reviewed literature as this is assumed to be between 10 and 100 years. Fifty per cent of the research considers lifetimes to be 50 years, where 19% considers this as 40 years and 9% considers this as 80–100 years. Lifetime considerations are important factors in life-cycle cost assessment studies, since these studies employ dynamic economic analyses with integrating uncertainty, sensitivity and risk assessments. Developments in material science and production technologies seem to increase the lifetime of the building elements.

Considerable attention on the design of the building elements for a long lifespan gives many innovative design solutions.

3.5 Conclusion

As today's world is facing severe environmental degradation, such as global warming and ozone depletion, the effort to reduce these problems through the accountability of every industry causing any adverse effects on the environment has been the major concern in recent decades. The concept of sustainability has been considered as a goal, a roadmap that can facilitate environmental and social improvements necessary for a future with conserved resources. While quantifiable aspects such as reduction of environmental impacts or energy consumption could largely be related with the built environment, conventional cost models were extensively used in order to assess the financial effects.

However, cost accounting for sustainability should not be a concept that can be evaluated independent from the environmental and social aspects of a sustainable built environment. In conventional project management processes economic performance is assessed through time, cost and quality (Abanda et al., 2013). Due to the rise of environmental concerns, the dimensions that were not included in the evaluation procedures of the buildings, such as environmental impacts, have become vital. The obligation for a more sustainable environment in order to reduce environmental impacts facilitated integration of life-cycle methodologies in evaluation procedures.

In this chapter, the aim has been to introduce cost modelling beyond the commonly employed static methodologies for building construction. In order to achieve such aim, the evaluation on how to include constraints such as time, energy performance and environmental impacts have been explored in depth within concepts of LCA methodology, which facilitates cost considerations as a vital decision-making criterion.

To facilitate investment decisions, it is necessary to depict economic interactions focused on the costs, which are incurred during the lifespan of a real-estate object and that must be borne by the investor, owner or operator.

Life-cycle costs could be divided into two cost areas, depending on the point in time that the costs are incurred. In addition, it is possible to divide life-cycle costs depending on their use. Thus, there are total investment, operating and additional costs. The concept of total investment costs is different from how the term is commonly used in the real-estate industry and does not represent all costs that are incurred during the project development. Rather, in this paper it is deduced that in the context of the life-cycle costing, investment costs represent all costs that increase the assets of the investor and must be capitalised.

To make an investment decision, economic interactions are illustrated by using investment calculation methods. The net present value and annuity methods, as well as the cost comparison method, are used primarily and fundamental differences are identified.

It is evident that with an increasing period of evaluation and a higher rate of interest on capital, the costs incurred in the context of the present value calculation become less important. At the same time, the uncertainty that results from the use of a large observation period of 30 years must be considered.

The result of the present value calculation contains too great a forecast of uncertainty when the observation period becomes too long. It can thus be stated that a life-cycle cost forecast that relates to the entire life cycle is not sensible. Rather, the observation periods for the present value method must relate to the useful economic life. This might be referred to as a useful cost forecast. In the context of another research project, which observation period is suitable depending on the type of real-estate object under consideration can be analysed.

Life-cycle cost assessment is a methodology that might significantly contribute to sustaining resources, both physically and financially. It has the potential to contribute to high construction productivity and may bring the opportunity to render profuse consumption of resources during design, investment, installation and management phases of a building. Simultaneous optimisation of investments, resources and environmental impacts would give the chance to obtain a positive impact on a sustainable future.

The next chapter focuses on the sustainable building process in three main phases, namely: pre-construction, construction and post-construction, outlining key aspects of each phase.

Notes

1   The condition where a commercial activity yields an accountable financial profit or gain.
2   Any system or process that must be analysed using probability theory is stochastic at least in part.

References

Accounting and sustainable development

Bebbington, J. and Larrinaga, C. (2014). Accounting and sustainable development: An exploration. Accounting, Organizations and Society, 39(6), 395–413.

Bebbington, J., Brown, J. and Frame, B. (2007). Accounting Technologies and Sustainability Assessment Models. Ecological Economics, 61(2–3), 224–236.

Frame, B. and Cavanagh, J. (2009). Experiences of Sustainability Assessment: An awkward adolescence. Accounting Forum, 33(3), 195–208.

Jasch, C. (2003). The Use of Environmental Management Accounting (EMA) for Identifying Environmental Costs. Journal of Cleaner Production, 11(6), 667–676.

Laitinen, E. K. (2014, June). Influence of Cost Accounting Change on Performance of Manufacturing Firms. Advances in Accounting, 30(1), 230–240.

Nigam, B. M. and Jain, I. C. (2001). Cost Accounting – An Introduction. Delhi: Prentice Hall of India.

Directives, guidelines and standards

BMVBS. (2013). Leitfaden Nachhaltiges Bauen. Retrieved from http://www.nachhaltigesbauen.de/fileadmin/pdf/Leitfaden_2013/Leitfaden_Nachhaltiges_Bauen_300DPI_141117.pdf

DGNB (2009). Das Deutsche Gütesiegel Nachhaltiges Bauen.

DIN 276-1 (2008). Building Costs – Part 1: Building Construction. Standard by Deutsches Institut Fur Normung E.V. (German National Standard).

DIN EN 60300 (2005). Dependability Management – Part 3-3: Application Guide – Life cycle costing DIN EN 60300-3-3. CENELEC.

EN 15804 (2012). CEN, Sustainability of Construction Works – Environmental Product Declarations—Core Rules for the Product Category of Construction Products. Brussels, Belgium: Comité Européen de Normalisation.

EPBD (2010). Directive 2010/31/EU of the European Parliament and of the Council on the Energy Performance of Buildings. European Union.

Fuller, S. (2014). Life-Cycle Cost Analysis (LCCA). Retrieved from Whole Building Design Guide. http://www.wbdg.org/resources/lcca.php (accessed 30 June 2015).

GEFMA (2005). Richtlinie 200 (2005a). Kosten im Facility Management.

ISO14040 (2006). Environmental management – Life cycle assessment – Principles and framework.

prEN15459 (2006). Energy efficiency for buildings – Standard economic evaluation procedure for energy systems in buildings. European Standard.

Facility management

Braun, H. (2004). Facility Management: Erfolg in der Immobilienwirtschaft, 4th edition. Berlin: Springer Verlag.

Hellerforth, M. (2006). Handbuch Facility Management für Immobilienunternehmen. Springer.

Kyrein, R. (2002). Immobilien – Projektmanagement, Projektentwicklung und –steuerung – 2. Auflage. Cologne: Müller.

Lai, J.H. and Yik, F.W. (2011). An Analytical Method to Evaluate Facility Management Services for Residential Buildings. Building and Environment, 46(1), 165–175.

Nävy, J. (2006). Facility Management: Grundlagen, Computerunterstützung, Systemeinführung, Anwendungsbeispiele – 4. aktualisierte und ergänzte Auflage. Berlin: Springer Verlag.

Life cycle assessment

Antipova, E., Boer, D., Guillén-Gosálbez, G., Cabeza, L.F. and Jiménez, L. (2014). Multi-Objective Optimization Coupled with Life Cycle Assessment for Retrofitting Buildings. Energy and Buildings, 82, 92–99.

Azari, R. (2014). Integrated Energy and Environmental Life Cycle Assessment of Office Building Envelopes. Energy and Buildings, 82, 156–162.

Cabeza, L.F., Rincón, L., Vilariño, V., Pérez, G. and Castell, A. (2014). Life Cycle Assessment (LCA) and Life Cycle Energy Analysis (LCEA) of Buildings and the Building Sector: A review. Renewable and Sustainable Energy Reviews, 29, 394–416.

Ding, G.K. (2008). Sustainable Construction – The role of environmental assessment tools. Journal of Environmental Management, 86(3), 451–464.

Gustavsson, L. and Joelsson, A. (2010). Life Cycle Primary Energy Analysis of Residential Buildings. Energy and Buildings, 42(2), 210–220.

Karimpour, M., Belusko, M., Xing, K. and Bruno, F. (2014). Minimising the Life Cycle Energy of Buildings: Review and analysis. Building and Environment, 73, 106–114.

Loftness, V., Hartkopf, V., Gurtekin, B., Hansen, D. and Hitchcock, R. (2003). Linking Energy to Health and Productivity in the Built Environment. Greenbuild Conference – Centre for Building Performance and Diagnostics, Cargenie Melon, Pittsburgh, PA.

Nässén, J., Hedenus, F., Karlsson, S. and Holmberg, J. (2012). Concrete vs. Wood in Buildings – An energy system approach. Building and Environment, 51, 361–369.

Ortiz, O., Castells, F. and Sonnemann, G. (2009, January). Sustainability in the Construction Industry. A review of recent developments based on LCA. Construction and Building Materials, 23(1), 28–39.

Pajchrowski, G., Noskowiak, A., Lewandowska, A. and Strykowski, W. (2014). Materials Composition or Energy Characteristic – What is more important in environmental life cycle of buildings? Building and Environment, 72, 15–27.

Pelzeter, A. (2006). Lebenszykluskosten von Immobilien: Einfluss von Lage, Gestaltung und Umwelt. Cologne: Müller Verlag.

Ribeiro, I., Peças, P., Silva, A. and Henriques, E. (2008). Life Cycle Engineering Methodology Applied to Material Selection, a Fender Case Study. Journal of Cleaner Production, 16(17), 1887–1899.

Rodrigues, C. and Freire, F. (2014). Integrated Life-Cycle Assessment and Thermal Dynamic Simulation of Alternative Scenarios for the Roof Retrofit of a House. Building and Environment, 81, 204–215.

Singh, A., Berghorn, G., Joshi, S. and Syal, M. (2011). Review of Life-Cycle Assessment Applications in Building Construction. Journal of Architectural Engineering, 17(1), 15–23.

Srinivasan, R., Ingwersen, W., Trucco, C., Ries, R. and Campbell, D. (2014). Comparison of Energy-Based Indicators used in Life Cycle Assessment Tools for Buildings. Building and Environment, 79, 138–151.

Verbeeck, G. and Hens, H. (2010). Life Cycle Inventory of Buildings: A contribution analysis. Building and Environment, 45(4), 964–967.

Watson, P. and Jones, D. (2005). Redefining Life Cycle for a Building Sustainability Assessment Framework. Sustainability Measures for Decision-support Fourth Australian Life Cycle Assessment Conference, Sydney.

Life-cycle cost assessment

Abanda, F.H., Tah, J.H. and Cheung, F.K. (2013). Mathematical Modelling of Embodied Energy, Greenhouse Gases, Waste, Time–Cost Parameters of Building Projects: A review. Building and Environment, 59, 23–37.

Brás, A., Gonçalves, F. and Faustino, P. (2014). Economic Evaluation of the Energy Consumption and Thermal Passive Performance of Portuguese Dwellings. Energy and Buildings, 76, 304–315.

Chel, A. and Tiwari, G. (2009). Performance Evaluation and Life Cycle Cost Analysis of Earth to Air Heat Exchanger Integrated with Adobe Building for New Delhi Composite Climate. Energy and Buildings, 41(1), 56–66.

Corgnati, S.P., Fabrizio, E., Filippi, M. and Monetti, V. (2013). Reference Buildings for Cost Optimal Analysis: Method of definition and application. Applied Energy, 102, 983–993.

Ebert, T. (2010). Zertifizierungssystem für Gebäude: Nachhaltigkeit bewerten, Internationaler Systemvergleich, Zertifizierung und Ökonomie. Munich: Aprinta Druck.

Fabrycky, W. J. and Blanchard, B. (1991). Life-Cycle Cost and Economic Analysis. Prentice Hall.

Finnveden, G. and Moberg, Å. (2005). Environmental Systems Analysis Tools – An overview. Journal of Cleaner Production, 13(12), 1165–1173.

Gluch, P. and Baumann, H. (2004). The Life Cycle Costing (LCC) Approach: A conceptual discussion of its usefulness for environmental decision-making. Building and Environment, 39(5), 571–580.

Han, G., Srebric, J. and Enache-Pomme, E. (2014). Variability of Optimal Solutions for Building Components Based on Comprehensive Life Cycle Cost Analysis. Energy and Buildings, 79, 223–231.

Hoffart, C. and Hirsch, T. (2011). Approach for Successful LCC Data Collection and Analysis. In: M. Singh, R. B. Rao and J. P. Liyanage (eds.), Proceedings of the 24th International Congress on Condition Monitoring and Diagnostics Engineering Management, 30 May–1 June, 2011, Stavanger, Norway, UK: Comadem International.

Horngren, C.T., Foster, G. and Datar, S.M. (1999). Cost Accounting: A Managerial Emphasis, 10th edition. Upper Saddle River, NJ: Prentice Hall.

Jacob, A. (1994). Basiswissen Investition und Finanzierung. Finanzmanagement in Theorie und Praxis. Dr. Th. Gabler Verlag; Auflage.

Möller, D. (2001). Planungs und Bauökonomie. Munich: Oldenbourg Verlag.

Pfarr, K. (1976). Handbuch der kostenbewußten Bauplanung. Wuppertal.

Pikas, E., Thalfeldt, M. and Kurnitski, J. (2014). Cost Optimal and Nearly Zero Energy Building Solutions for Office Buildings. Energy and Buildings, 74, 30–42.

Reports

Brundtland (1987). Report: Our Common Future. UN Commission.

EPA (2014). Chapter 4: Life Cycle Impact Assessment. Retrieved from United States Environmental Protection Agency. http://www.epa.gov/nrmrl/std/lca/pdfs/chapter4lca101.pdf (accessed 6 July 2015).

UN (1992). Report of the United Nations Conference on Environment and Development. Rio de Janeiro.

Sustainable architecture

Guy, S. and Farmer, G. (2001). Reinterpreting Sustainable Architecture: The Place of Technology. Journal of Architectural Education, 54(3), 140–148.

4 The Sustainable Building Process

Begum Sertyesilisik

4.1 Introduction

The highly competitive environment of the construction industry forces construction companies to manage construction projects effectively and efficiently, and increases the importance of customer satisfaction and value creation. Construction companies need to scan the business environment to analyse opportunities and potential rewards. The demand and need for sustainable buildings has increased, especially as global warming has highlighted their advantages. Construction companies need to adapt to these changes in the construction market. Built environment and construction activities need to be more sustainable (Hill and Bowen, 1997; Barrett et al., 1999; Holmes and Hudson, 2000; Morel et al., 2001; Scheuer et al., 2003; Cole, 1999; Ding, 2005). Environmental matters need to be incorporated into the decision framework of project appraisal at an early stage in order to achieve sustainable construction (Ding, 2005). Being green and sustainable needs to become a way of doing business and integrated into the contractor's corporate culture so that contractors can ensure an efficient and effective construction process, using renewable resources and reducing the amount of resources used and waste generated (Glavinich, 2008: 2–6). Sustainable construction, starting at the planning stage, continues throughout the building's life to its deconstruction and the recycling of resources (Hill and Bowen, 1997). Green design requirements affect all these stages, from the contractor's procurement process to the construction phase and finally the project closeout (Glavinich, 2008: 34). This chapter focuses on the sustainable building process in three main phases, namely: pre-construction, construction and post-construction.

4.2 Pre-construction phase

The pre-construction phase is the first phase of the sustainable building process. This phase includes: contracting for sustainable construction, design, establishment of the team and determination of site management principles for the specific project.

4.2.1 Contracting for sustainable construction

Contracts are legally binding agreements between the contracting parties describing their rights and obligations and scope of the work. Contracts allocate risk among the contracting parties affecting the cost of the work. Contracts are the main basis for the relationship of the contracting parties. For this reason, well-prepared contracts can help the parties to reduce the risk of dispute whereas contracts prepared without sufficient care can cause disputes. Contracting for sustainable construction is for the most part similar to contracting for any other project, except the owner's desire to achieve a certain level of green or sustainable building performance or certification that affects the work's scope (the subject of the contract), the construction process, the contractor's costs, schedule, productivity and the procurement process, the construction and the project closeout and how the contractor and its subcontractors carry out the work at the project site (Glavinich, 2008: 33–34, 59). Project delivery system, contract documents, contract conditions, specifications, drawings, addenda and subcontract are all analysed in the following paragraphs within the scope of contracting for green construction.

Project delivery method

The project delivery method influences the contractual relationship between the stakeholders (Lam et al., 2010b) and the contractor's role, responsibilities and risks (Glavinich, 2008). Project delivery processes for sustainable projects are generally more complex and have more stakeholder interactions than project delivery processes for non-green construction (Hill and Bowen, 1997; Reed and Gordon, 2000; Lapinski et al., 2006). The range of procurement systems has proliferated due to the increasing complexity of projects and the need to manage costs more efficiently (Masterman, 2002: 34). There are three main project delivery methods (ASHRAE, 2006: 83): the construction manager approach, the design-bid-build (traditional) approach and the design-build approach. These methods are briefly explained next, emphasising the advantages and disadvantages of each method for sustainable projects:

- In the construction management project delivery method, the construction manager supports the owner during the sustainable design process through constructability reviews, value analyses and budget and schedule reviews, so that waste is reduced and the sustainability of the construction process and its output is enhanced (Glavinich, 2008: 38–39). This method enables extensive co-ordination among the consultants and specialist contractors. For large construction projects requiring complex and innovative technologies, there is a need for single point construction management (Peace and Bennett, 2003a). All of these factors reveal the suitability of this method to sustainable projects as they need collaborative work and careful control.
- In the design-bid-build project delivery method, following the accomplishment of the design by the design team or by the design company, the owner contracts the general contractor to carry out the construction (Glavinich, 2008: 36). This separation of design and construction often leads to issues of buildability on site (Dow et al., 2009: 102) causing the need for re-work, which consequently increases the amount of wasted material. As the contractor is not integrated into the design phase, the owner cannot get the benefit from the contractor's expertise with respect to value analyses and constructability reviews during the design phase (Glavinich, 2008: 37). This procurement method is slow, expensive and can cause clients to become dissatisfied with its results (Masterman, 2002). This method is not advisable for sustainable projects due to the lack of involvement of the contractor in the design

and construction phases, as sustainable projects require integrated teamwork due to their interdisciplinary and relatively complex nature.

- In the design-build project delivery method, one contract covers both design and construction. Following an initial client brief, possibly accompanied by an outline design, the design-build contractor will employ all the consultants and specialist contractors needed to produce the building as subcontractors (Dow et al., 2009: 102). In this scenario, as the contractor is solely responsible for the project, he/she needs to analyse the owner's requirements to ensure that it is possible to meet the project criteria while complying with sustainability performance and certification aims (Glavinich, 2008: 71). The design-build project delivery system allows the owner to benefit from construction expertise during the design phase and from design expertise during the construction phase (Glavinich, 2008: 40). This method outperforms the design-bid-build method with respect to most aspects of project performance (i.e. cost predictability, overall speed) (Peace and Bennett, 2003b).

Contract documents

Well-prepared contracts can help the contracting parties to execute construction successfully, reducing the risk of dispute. This, in turn, helps to protect the relationship between the contracting parties and, in all likelihood, improves customer satisfaction, positively affecting possible future work opportunities for the contractor. The contracts need to be carefully drafted taking into consideration the following main points, which include but are not limited to:

- paying attention to the clarity of language and clauses to avoid any potential misinterpretation
- describing the work scope clearly (since if this is not clear it can cause conflicts in determining the amount of any extension in time and payment needed, especially in the event of variations and change orders)
- fair allocation of risk among the contracting parties
- order of contract documents, helping to clarify ambiguous or conflicting statements among various documents
- describing variation and change in order procedures
- taking into account main dispute areas, which can be grouped as financial, temporal, compliance, technical documents and clause related issues (Sertyesilisik, 2010)
- describing time-related topics, especially in the following clauses: determination of any extension of time needed, liquidated damages, duration of the work, effective date of the contract and *force majeure* conditions
- describing payment-related topics with particular attention on the following points: payment amount, payment type (i.e. lump sum, unit price), payment conditions and time and delay in payment.

The contractor needs to review the contract documents in detail with respect to his/her rights and obligations, risk allocation, legal aspects, financial and technical feasibility, as well as sustainable building requirements in detail. The sustainable project requirements can be described in different parts of the contract documents (i.e. in the project specifications and/or referencing to the sustainable building certification) (Glavinich, 2008: 44). There is a need, in construction work in particular, for the following documents to be prepared: contract conditions, specifications, drawings, addenda and subcontracts.

Contract Conditions: These consist of: general conditions; supplemental and special conditions (dealing with the owner's special requirements and restrictions for the project work) and combined supplemental and special conditions. Sustainable construction requirements can be covered in various parts of the contract, but particularly under the supplemental conditions (which may be combined with the special conditions) (Glavinich, 2008: 66–67).

Specifications: The technical aspects of the contracted works are for the most part explained in the specifications. For this reason, it is important that the specifications are carefully drafted eliminating unclear statements, technically impossible or contradictory requirements. The specifications are an important management tool to enhance the sustainability of construction resources (Lam et al., 2010b). The contract can contain technical specifications and/or performance specifications. Technical specifications describe the technical aspects and all details (i.e. brand, colour) of the material/equipment to be used in the building. Performance specifications, on the other hand, provide the performance of the space, room or building (i.e. acoustics performance, sustainability performance). Sustainability performance of the building can be described as performance specification, referring to the specific building certificate and its level. The contractors need to avoid mixed specifications that provide both technical and performance aspects as they can contradict each other, putting the contractor in a difficult position (Glavinich, 2008: 48). For example, conforming to the technical specifications may not result in successful accomplishment of the performance specifications. Once the contract is signed, the contractor becomes liable to accomplish the contract requirements, as a contractor is expected to be technically competent and to be able to foresee any errors, or contradictory requirements in the contract. In the event of a mixed specification having conflicting technical requirements, the contractor should ask the owner for clarification during the bidding phase. The addenda, the clarifications made by the owner upon the contractors' requests during the tendering phase, become part of the contract document, enabling the contractors to bid more accurately. Following the introduction of various green building assessments in different regions based on local characteristics, many organisations that provide a specification service (i.e. BRE's (2010) "The Green Guide to Specification", which focuses on the environmental impacts of building materials, or the assessment criteria by the California Energy Commission (2007), the green building specification, which focuses on energy efficiency and environmental sustainability) have already published or are planning to develop green specifications for these green building assessments (Lam et al., 2010b). Emphasising the need for the introduction of green specifications to advance green performance in construction through contract management, Lam et al. (2010a) defined "green specifications" as the written instructions for construction practice relating to the use of materials and working procedures to ensure sustainability of development in terms of economics, community, technical feasibility and the environment with both global and local considerations.

Drawings: As the drawings can include sustainable project requirements or relevant references, in design-bid-build, the contractor needs to carefully examine the design provided by the owner with respect to conflicting goals or technically impossible requirements and inform the owner before the bidding phase. In the design-build procurement model on the other hand, the employer's requirements document provided by the owner needs to be analysed with respect to conflicting technical and performance requirements as well as technically impossible require-ments. The contractor needs to ensure that the design, based on the employer's requirements document, is in compliance with the owner's expectations and sus-tainable building goals stated in the contract documents.

Addenda: Addenda are explanations to the questions of the participants of tender and modifications to the contract documents, during the bid period. Addenda become part of the contract and need to be taken into consideration when bidding, as changes or explanations to the original contract documents can affect the contractor's understanding of his rights and responsibilities affecting the tendering cost.

Subcontracts: As the contractor is responsible for the performance of all subcontractors, the contractor needs to prepare well and ensure subcontracts are clearly written to take into account work scope, clarity and consistency in all contract tiers. All contract clauses need to be clearly written to avoid misun-derstandings on the rights and obligations of the contracting parties. Clarity is especially important in payment-related clauses (amount of payment, payment time, delay in payments, progress payments, liquidated damages etc.) and time-related clauses (duration of the work, deadline, approval timings and conditions, extension of time, *force majeure* etc.). The main contract and the subcontracts are recommended to be in the same currency in order to avoid parity difference originated losses. Furthermore, the work scope should be clearly defined in each subcontract. It should not overlap with other subcontracts for the project. The rights and obligations should be clearly defined to avoid confusion. As varia-tions and change orders can affect the work scope, variations and change order procedures, as well as parties' rights and obligations under such circumstances, should be clearly defined. Variations in work scope can result in additional work causing need for additional time and payment. Sustainable project requirements and necessary submittals need to be incorporated into subcontracts and sup-ply agreements without gaps or overlaps in the subcontractors' work scopes ensuring consistency among contracts in each contract tier (Glavinich, 2008: 63–122). For example, the main and subcontractors should specify an on-site construction waste sorting requirement in their contracts as, according to Poon et al. (2001), the majority of contractors would not perform on-site construc-tion waste sorting unless it is specified in the contract. Subcontractors should train their workers on safety as well as on sustainability topics and they should take the necessary safety precautions on site as well as sort on-site construction waste. All assigned or approved subcontractors need to be competent in dealing with the sustainable building construction requirements. The contractor should pay attention to the sustainability requirements in its request for bid or proposal to subcontractors so that the subcontractor can incorporate them into its bid (Glavinich, 2008: 115). In the case of the design-build-procurement method, if the contractor cannot carry out the design in-house, there is a need for the contractor to subcontract a design company based on the main points raised previously.

Lawyers' support should be sought in analysing the main and subcontracts with respect to legal issues whereas staff with technical backgrounds should analyse the main and subcontracts with respect to technical aspects. Other documents, such as commissioning plans, can also be part of the contract.

4.2.2 Establishing the team

Establishing a competent team is important to project success; in other words, firms that manage to "attract, develop and retain top talent will thrive; those that do not will face significant struggles" (Holtom et al., 2005: 337). For this reason, establishing a team consisting of competent staff is an important start and the qualified team members should be retained at least throughout the project as employee turnover affects performance and competitiveness of companies (Jones et al., 2010: 271). Contracting parties' experience, expertise, skills, financial status and even personalities affect smooth and successful management of the project. For example, owners' financial problems (Clough and Sears, 1994; O'Brien, 1998); skills shortage (Arain et al., 2004); obstinate nature of owner (Wang, 2000; Arain et al., 2004) and poor workmanship (Fisk, 1997; O'Brien, 1998) due to lack of knowledge of the workers can prevent the smooth execution of the project affecting the contracting parties' relationships. The criteria that should be considered for the appointment of contractors, subcontractors and designers are outlined next:

- **Contractor Appointment:** Appointment of the contractor plays an important role in the project success. For this reason, the owner needs to choose the most appropriate tendering type conforming to his expectations. Tendering types are briefly explained in the following paragraphs:

    o In the case of open tendering, there is no restriction on companies taking part in the tendering process. This might result in the company proposing the lowest price for undertaking the construction. However, it might not be possible to do the construction work for the lowest price agreed, as the appointed construction company may not be competent in undertaking the work scope defined in the contract. For this reason, as green and sustainable projects are more complicated compared with non-green ones, open tendering is risky.

    o Selective tendering, on the other hand, helps to overcome the risk of the appointment of a non-qualified contractor, due to the fact that it is based on the pre-selection of the companies willing to participate in the tender. For example, if only qualified contractors participate in the tender, the winning contractor will be one of the qualified contractors, enhancing the project success. For this reason, it is valuable to preselect the teams or contractors from whom pricing is requested (ASHRAE, 2006: 85). A contractor for a sustainable building project can be preselected based on price, experience of working on green building projects and the knowledge and capabilities of the project team, or based on best-value selection, which involves a combination of the previous points (Glavinich, 2008: 42–43). Experience in similar work, record of past performance by responsible references, financial capability and workload of potential constructors should be analysed in the prequalification process (ASHRAE, 2006: 84).

○ Serial tendering is based on collaboration between the contracting parties for a series of contracts in the future assuming the contractor performs satisfactorily in the awarded contract. If the contractor is motivated because of the prospect of undertaking future work, the chances of the project succeeding are improved.

○ Competition-based tendering enables the owner to select the design that best complies with the owner's requirements from among those submitted by the companies participating in the competition. The disadvantage of this type of tendering is that the competent contractors or design companies are reluctant to participate in competitive tendering, due to the cost of the preparation of design, which increases their overhead expenses. Each unsuccessful bid reduces their competitiveness in other tenders.

- **Subcontractor Appointment:** The contractor is responsible for the performance of all subcontractors even if they are appointed or approved. In cases where the owner has appointed subcontractors who perform unsatisfactorily, the contractor can notify the owner accordingly based on objective criteria so that another subcontractor is appointed. The owner might agree or disagree with this notification. Since the successful accomplishment of a sustainable construction project is dependent on the subcontractors' satisfactory performance, it is essential that the contractor discusses the green or sustainable project requirements with the subcontractors, as well as educates them about their role and the project's green or sustainable criteria and assists them in their training (Glavinich, 2008: 111–112, 124–125). Al-Khaddar et al.'s (2012: 126) study revealed that a "deep learning technique" should be used when training staff so that they embrace and use "green issues" in their everyday practice, leading to a continuous improvement in green practice. The deep learning technique is based on understanding the essence of the topic. Deep learners, in their learning process, look for meaning relating the new and the previously taught information together and linking them to real life scenarios. Deep learners fully engage with the topic and fully understand it (Warburton, 2003) through holistic acquisition of knowledge and tracing out possible connections between concepts, facts, events and phenomena (Ramsden et al., 1989). For this reason, a deep learning approach can result in good understanding and long-term information retention (Al-Khaddar et al., 2012: 128–129). The contractor needs to request and assess subcontractor bids or proposals taking into account the selection criteria based on project scope and qualifications needed (Glavinich, 2008: 118). In order to harmonise the construction works, the main contractor should provide the overall construction work schedule so that different subcontractors working in different phases can effectively perform their work on time.

- **Designer/Design Company Appointment:** The designer/design company should be appointed based on the project scope, the designer's/design company's experience and expertise in the fields of project scope and sustainability design principles. The expertise of the designer/design company can be understood through analysing past projects that have been undertaken by the designer/design company. The expertise of the design company in the relevant field can be understood by analysing the "state-of-the-art" of the company with respect to human resources and infrastructure, including the database for past projects. In the event that the company is not an organisation with good training/handover plans in place (this situation occurs mainly where the experience gained by certain staff members is not written down and transferred into the company's database), then there is a high probability that any experience gained will be lost when the relevant staff leave the company. Accordingly, a company that has experience in a specific field (i.e. sustainable design) may not in fact possess the expected expertise. For this reason, the designer/design company who carries out the design needs to have a strong background in sustainable design and construction, including the sustainable building certificate to be used in the project. In the design-build procurement method, the contractor needs to define the expertise and scope of design services needed so that he can start drafting a request for qualifications (Glavinich, 2008: 87).

4.2.3 Pre-design consideration and design development

The design team needs to prepare a design based on project objectives taking into consideration value engineering practices. The more detailed and accurate the design is, the more accurately the contractors can bid in the design-bid-build, reducing the need for variation and change orders during the execution phase of the construction. Taking into account value engineering practices in the design process can help with the inclusion of green and sustainable construction principles. Sustainable design should enable the creation of value for the environment both now and for future generations as well. As stated in Keane et al. (2010); design complexity (Fisk, 1997; Arain et al., 2004); failure to include value engineering in design phase (Dell'Isola, 1982); poor working drawing details (Geok, 2002; Arain et al., 2004) and inadequate project objectives (Ibbs and Allen, 1995) can all increase the need for variations and change orders.

A sustainable project requires a schematic design effort with co-ordination among various groups including construction professionals, mechanical engineers, facilities managers, building occupants and utility companies (Koltz and Horman, 2010: 595). For this reason, sustainable projects are designed in an integrated way. Traditional design process has a linear progression and a strict hierarchy of communication, with minimal interaction between the design team members and a front-end loaded cash flow of fees, whereas an integrated design process is a more iterative process providing flexibility and dynamism in the engagement of all team members so that there is scope for ongoing learning and the capacity to address emergent features and strategies [WBDG (Whole Building Design Guide) and MFE (Ministry for the Environment) websites]. As in the traditional design process the design team's input peaks during detailed design and reduces during the construction phase, so the final design may not incorporate all design objectives leading to problems during the operational phase of the building (MFE website). Integrated design, with the participation of all stakeholders in the process, helps to produce far superior buildings than the conventional relationships with precisely the same green brief for a building project (Kibert, 2007: 600). The integrated design process (ASHRAE, 2006: 87) ensures that as many of the interested parties (architects, engineers, costing specialists, operations people and other relevant people including design facilitators, energy specialists and subject specialists) as possible are represented on the design team as early as possible, enabling interdisciplinary work from the beginning of the design process. As the design team is involved in all processes, this risk of problem emergence decreases. Discussion and documentation by the owners and the design team of the relative importance of various performance and cost issues is critical. "IWBDP produces: higher-performance buildings; a better indoor environmental quality (IEQ); a team spirit among designers and the client" (Masstech website). If facilitated effectively, integrated design can enable the team to generate ideas greater than the initial sum of the team parts.

Sustainable and integrated design requires clear and measurable sustainable project criteria as prioritising goals helps the designer and client to understand important or discardable goals and the flexibility of the proposed solutions (Architectural Research Centers Consortium, 2007: 191–192). Furthermore, sustainable and integrated design requires highly motivated and committed owners and project team members (ASHRAE, 2006: 6–7) supported by clear communication among them (Busby et al., 2007), strong leadership (ASHRAE, 2006: 82) and skillful management (i.e. selection of design team members with experience in integrated design). Clear communication can be enhanced through clear communication protocols adhered to by the project team (WBDG website).

Sustainable and integrated design requires an iterative process with feedback cycles (Busby et al., 2007) as this iterative process enables design alternatives to be evaluated, refined, evolved and optimised to come up with the most effective combination (ASHRAE, 2006: 87). Throughout this iterative process, there is a need to look at the long-term environmental impacts of the building (ASHRAE, 2006: 74) considering: life-cycle cost (ASHRAE, 2006: 81; Masstech website) recyclability aspects, energy-saving features and minimal impact on natural resources (Lund, 2001) as well as the environmental consequences of the structure's design, orientation, impact on the landscape and materials (Kim and Rigdon, 1998: 12). The interdisciplinary design team needs to think ecologically about design, minimising environmentally destructive impacts (Cowan and van der Ryn, 1996) and taking into account the fact that building design can help to reduce waste (Chung and Lo, 2003: 125). Key indoor environmental quality features, water efficiency systems, renewable energy systems, site design and core lighting design need to be included in the project (Yudelson, 2007).

Sustainable and integrated design can be supported, in the pre-design phase, through an interactive design charrette with the participation of all related stakeholders. This charrette is a focused and collaborative brainstorming process allowing various stakeholders to make their priorities and concerns known (Masstech website). The "eco-charrette" concept can be defined as follows (the Betterbricks website):

> "Eco-charrette, in other words sustainable design or environmental design charrette, is an intense meeting, half a day or more, in which all participants in a building design project focus on ideas for efficient use of energy and resources, sustainable development goals, strategies and integrated design solutions in the new building. The group generates goals and then develops strategies for accomplishing those goals…"

Sustainable and integrated design requires a holistic and integrated approach to design aimed at restoring the balance between the natural and built environments (Drager, 1996; du Plessis et al., 2002; Ross et al., 2010: 435). Holistic thinking is needed for integrated solutions, easy optimisation of design, easy incorporation of sustainable features and cost-effective design. As the whole team is involved from the beginning of the design phase, a sustainable and integrated design process takes longer, requires greater co-ordination among the team members and relies more on front-end loaded payment than non-green designs (Glavinich, 2008: 81).

To ensure a sustainable and integrated design, sub-consultants should be involved in the pre-design activities. This helps to prevent any costly changes that may otherwise occur at a later stage (WBDG website). These pre-design activities can include: performing a cost estimate (which could be part of the selection process for a builder) and considering other prerequisites such as bonding capacity, experience and references (WBDG website).

The design team, who should be involved throughout construction, commissioning and even post-occupancy (Masstech website), can provide input into the initial OPR document to help establish sustainable/green project criteria (ASHRAE, 2006: 82). To achieve a sustainable/green design, they should follow three main steps (ASHRAE, 2006: 88): reduce the loads (cooling load, solar loads, lighting loads and heating load); apply the most efficient systems and look for synergies. The designers should regard very early contact with a potential owners as an opportunity to steer the project in a green direction (ASHRAE, 2006: 7), implementing regular design reviews (internal design reviews and owner design reviews) to check that the design is progressing as planned with regard to the owner's requirements (Glavinich, 2008: 100). The design team can also contribute to increasing recycling rates by considering the regional capacities for recycling, the materials' lifetime and component sizes and by avoiding the use of too many different types of materials and materials that become obsolete (Srour et al., 2010). These decisions can increase and improve recycling during the operation, maintenance and demolition phases (Srour et al., 2010).

Commissioning aspects need to be considered in the sustainable and integrated design phase (Masstech website) through incorporation of a commissioning agent into the building team in the pre-design phase (Yudelson, 2007). The pre-design phase plays an important role in the success of the design and construction phases, enabling the corrective actions to be taken effectively (ASHRAE, 2006: 42). The commissioning phase needs to be considered in the pre-design and design phases as explained by ASHRAE (2006: 42) in the following statements:

> "In the pre-design phase, the commissioning process begins with development of the owner's project requirements. In the design phase, the commissioning authority provides checklists to designers to assist them in their design quality control process and alert the designers on the specifics the commissioning authority will be focusing on during the commissioning design reviews" (ASHRAE, 2006: 42).

During the sustainable and integrated design phase, life-cycle management can be applied to the whole construction process so that the sustainability performance of the building is improved (i.e. a proper design and choice of building materials during the pre-construction phases can improve the energy efficiency during the operation phase and the final distribution of buildings' consumption for heating and cooling) (Ortiz et al., 2009).

4.2.4 Construction site arrangements

Construction site arrangements can be analysed under two subheadings, namely: site assessment and site waste management.

Site assessment

Careful site selection can minimise the negative environmental impacts of the project, from construction to the occupation phase and it can lower first cost, operating and maintenance costs, environmental cost and people cost (ASHRAE, 2006: 55). Sustainable site planning requires the assessment of the building site, which is done at three levels (ICAEN, 2004: 17–18):

- Site selection is about identifying and weighing up the appropriateness of the site with respect to sustainable building design criteria (ICAEN, 2004). Considerations for site selection include:

  - transportation of materials and labour to construct the project; loss of land that supports biodiversity; the highways, roads and bridges required to provide access to the facility; the infrastructure needed to support operation of the facility and the proximity of the facility to residential and other services for building occupants to reduce natural resources needed for transportation; building form and orientation, nearby pollution sources, ambient air quality, groundwater levels, site drainage, availability of or access to various energy sources (including renewables) (ASHRAE, 2006: 55–56)
  - the mitigation of air, land, water and noise pollution both during construction and during building use and light pollution from buildings; transport-related issues including the provisions for car parking, alternative transport (e.g. facilities for cyclists) and providing suitable access to buildings to avoid congestion and reduce nuisance (Brunett, 2007: 35).

- Site analysis is the evaluation of all the on- and off-site determinants (i.e. environmental, cultural, historical, urban or infrastructural affecting the development of the site and its building programme) to establish the site characteristics so that inputs necessary for building design parameters are gathered (ICAEN, 2004).
- Site development and layout affect the efficiency and safety of the construction work as well as material which might end up wasted and the rate of recycling (ICAEN, 2004)

Waste management

Construction waste is increasingly becoming liable to contamination by hazardous substances due to increased use of organic polymers and chemical additives in construction (Lahner and Brunner, 1994). Reducing the volume of waste disposed as landfill is the primary objective of a waste management programme (Lingard et al., 2001: 813). Waste minimisation in construction begins with the design phase, continues throughout the construction phase via site waste management and ends in the post-construction phase with the demolishing of the construction when the demolished building materials are reused or recycled. In the pre-construction phase, a waste management plan needs to be prepared before the project has started to identify the potential types of waste used on site in order to enable sustainability and reduction in waste (Sertyesilisik et al., 2012). A waste management plan can be prepared based on the BRE's SMARTWaste Plan as well. Agreements determining who is responsible for waste on-site should be carried out between contractor and sub-contractors (Sertyesilisik et al., 2012). Waste minimisation has environmental benefits including (Lingard et al., 2000: 384):

- prolonging the life of landfill sites and reducing primary resource requirements (CIRIA, 1993)
- reduction in transport needs and the associated impact on the environment
- social benefits including the avoidance of creating new and undesirable landfill sites
- an essential addition to the time-cost-quality imperatives of traditional construction management practice (Ofori, 1992; Hill and Bowen, 1997).

In addition to these benefits, the contractor does not only earn money from the scrap created by the construction activities on site, save on landfill taxes and decrease the land space used for landfills (Johnston and Mincks, 1995) but also improves their image with respect to social and environmental responsibility. Despite these benefits, there is a reluctance in the construction industry to take the necessary steps for achieving effective waste management especially due to the following main barriers (Lingard et al., 2000: 383–389): the construction industry's culture (Federle, 1993); conflicting goals of the participants in the construction industry; the emphasis on productivity goals and speed of construction; subcontractors' workers feeling that their efforts will go unrewarded (Lingard et al., 2001: 814); employees' perception that waste management is not cost effective (Lingard et al., 2000); the managers' perception that environmental issues are less important than cost, time or quality objectives (Lingard et al., 2000); pressures to complete work quickly leading to usage of components from new material instead of using previously cut pieces (Federle, 1993); operation costs, lack of trained staff and expertise and lack of government legal enforcement (Shen and Tam, 2002). An on-site sorting process can disturb the construction process and it can be difficult on sites in a city centre as it requires a vast amount of containers (Sertyesilisik et al., 2012).

Effective construction waste management depends on and can be improved via (Lingard et al., 2000: 383–390):

- changing the behaviour of individuals involved in the construction process through the implementation of waste management procedures and policies
- the organisation of work on site or within the company as a whole
- shared employee perceptions of the importance and rewards associated with company waste management objectives
- construction workers' involvement in waste management
- including waste management in occupational health and safety committee agenda items and/or integrating this topic into systems for dealing with employees' concerns

- policies addressing both site construction workers' and managers' concerns
- management's visible commitment to waste management
- effective meetings to increase the involvement of construction site workers in waste management issues.

Lingard et al. (2001: 814–815) recommend rewards for meeting waste management objectives. This ensures that workers understand the benefits of waste management and see goal setting and feedback as useful components of a good waste management programme.

4.3 Construction phase

Throughout the construction phase, the contractor needs to fulfil his contractual obligations and he needs to make sure that the accomplished construction is in compliance with the contract requirements. Efficient project management can enable the contractor to fulfil his obligations on time, within budget and in compliance with the contract documents satisfying the owner's expectations. Following the delivery of the construction site to the contractor, the contractor needs to perform mobilisation activities based on the site plan prepared, taking into account safe and efficient performance of the construction work. The important aspects in the construction phase include:

- Submittals required in the contract and in the targeted green certificate should be prepared and submitted on time.
- All correspondences among the contracting parties should be documented to protect rights. In the event of a request for an extension of time or for an additional payment, claims should be supported by relevant documents. A lawyer's support should be received especially in the event of claims. In the event of a dispute, the dispute resolution clause should be followed. The amicable settlement approach, however, should be the first step to protect the relationship among contracting parties without disrupting work flow.
- The green building certificates' requirement should be adhered to (more detail on the green building certificates is provided in Chapter 5).
- A construction programme should be prepared and distributed to the subcontractors so that they work in harmony.
- Procurement of the materials should be done in accordance with the work progress and programme. Storage and handling of the materials should be carefully done so that materials are not wasted.
- Meetings at different managerial levels are important for keeping communication among the contracting parties efficient.
- Quality control should be effectively carried out throughout construction so that defects are avoided.
- The commissioning authority should provide construction checklists to the contractors to assist them in their quality control process and to verify that the contractor's quality control process is working (ASHRAE, 2006: 42).

Efficient and effective execution of the construction is needed as claims, variation and change orders can arise due to various reasons including (Keane et al., 2010): fast-track construction (Fisk, 1997); unforeseen problems (Clough and Sears, 1994; O'Brien, 1998); poor co-ordination (Arain et al., 2004); differing site conditions and poor workmanship (Fisk, 1997; O'Brien, 1998); change in design (Fisk, 1997; Arain et al., 2004); desired profitability (O'Brien, 1998); conflicts among contract documents (CII, 1986); owner's financial problems (Clough and Sears, 1994; O'Brien, 1998) and contract related inadequate project objectives (Ibbs and Allen, 1995). As Keane et al.'s (2010: 92) research revealed all of these factors have an impact on cost (i.e. increase in overhead expenses (O'Brien, 1998); rework and demolition (CII, 1990; Clough and Sears, 1994); quality (i.e. quality degradation CII, 1994; Fisk, 1997); time (i.e. procurement delay (O'Brien, 1998); rework and demolition (CII, 1990; Clough and Sears, 1994); organisation and related effects on reputation (i.e. the tarnishing of a firm's reputation (Fisk, 1997); dispute among professionals (Fisk, 1997) and other effects (i.e. progress affected without delay CII, 1994). Furthermore, they have influence on environmental effects of the construction process. The environmental impact of construction needs to be minimised throughout by employing effective measures, that is, site waste management, noise, vibration and emission control and so on. In the building phase, the construction and operation processes should be examined for ways to reduce the environmental impact of resource consumption (i.e. minimise site impact and employ non-toxic materials) and the long-term health effects of the building environment on its occupants and the environmental impact of actual construction and operation processes should be considered (Kim and Rigdon, 1998: 12–24).

Contractors are responsible for recycling materials through protecting, handling and treating them properly and through using the most effective methods of recycling (Srour et al., 2010). Construction waste has great potential for recycling, recovery and re-use (Merry, 1990; Cosper et al., 1993; Schlauder and Brickner, 1993; von Stein and Savage, 1994; Lingard et al., 2001) mainly depending on motivational influences on the behaviour of construction workers (Baetz et al., 1991; Beaumont et al., 1993; Lingard et al., 2001: 809). Causes of waste in the construction phase can be classified into two groups; damaged material and unused material:

- Damage to material can occur "through mishandling, due to weather, delivery, theft of materials and vandalism, inadequate storage of materials; lack of recycling facilities on-site; poor layout of the site; poor information communicated between employees; over-ordering of materials with a high surplus material percentage; poor organisation by a site manager; casual attitudes undertaken by some sub-contractors resulting in high levels of waste and a lack of interest in recycling by skip hire companies for various reasons" (Sertyesilisik et al., 2012).
- Unused material can occur due to "unused products; off-cuts from cutting materials; waste from application processes; improper storing space and methods; weather conditions and delays in forwarding information on sizes of materials to be used." (Osmani et al., 2008: 1153).

These causes reveal the importance of the construction phase. Improving site practices and appropriate sorting at the generation source help to reduce waste (Chung and Lo, 2003: 125). Effective methods to achieve sustainability in the construction industry in the construction phase can be listed as follows (Sertyesilisik et al., 2012):

- collaboration with the waste companies that offer free waste skip disposal or usage of colour-coded bins to segregate waste for recycling;
- appointment of a designated waste manager to deal with the delivery and storage of materials;
- segregation of waste materials for recycling and/or to be collected by recycling contractors should be done by using separate, covered and labelled storage bins (e.g. different coloured bins for different materials) placed in the most convenient location to the site entrance and providing clear instructions and training to all employees on waste procedures. The general contractors are perceived as being more responsible for segregation of waste on site than the subcontractors;
- delivering and storage should be carried out taking into account delivery times and methods of packaging related to planned progress on-site. All materials should be supervised and recorded when off loaded. They should be locked up on site. The packaging should be checked and carefully stored in secured storage units with access only by the waste manager;
- Earth Exchange (EE), which enables you to find users for surplus construction materials and to acquire materials for reuse sourced from other construction projects or suppliers in the construction site area.

The waste can be either transported directly to sorting facilities with high charges or they can be processed through on-site sorting before disposal at landfill or public fillings (Hao et al., 2007). Hao et al. (2007) showed that the most effective means of waste management is on-site sorting of construction waste that involves inert waste used for land reclamation and non-inert waste disposed of in landfills.

4.4 Post-construction phase

After the building is fully operational, it is often useful to conduct a Facility Performance Evaluation to assess how the building meets the original and emerging requirements for its use (the WBDG Aesthetics Subcommittee, 2009).

Commissioning is the process of ensuring that systems are designed, installed, functionally tested and capable of being operated and maintained to perform in conformity with the design intent as defined in ASHRAE Guideline 1-1996 (Djuric and Novakovic, 2010: 510). The owner's project requirements are the foundation for both the commissioning process and for defining the objectives and criteria that will guide the project delivery team (ASHRAE, 2006: 41). As the commissioning process is a quality-oriented process for achieving, verifying and documenting that the performance of facilities, systems and assemblies meet defined objectives and criteria, commissioning is an essential part of green building design and construction through verification that the goals defined by the owner and integrated by the design and construction team are actually achieved as intended (ASHRAE, 2006: 41).

Once the construction has been successfully accomplished and accepted by the owner, the contractor starts the demobilisation phase repairing the environmental damage caused by construction activities. Following this, the construction itself can be utilised. Ortiz et al. (2009: 592) research concludes that in the whole construction process, the operation phase is the most critical phase because of the higher environmental burden emitted into the atmosphere.

In the post-building phase, after the economic life of the building, the environmental consequences of structures that have outlived their usefulness are examined to understand the most suitable solution out of three possibilities: reuse and recycling of components (allowing a building to become a resource for new buildings or consumer goods) and disposal (requiring incineration or landfill dumping, contributing to an already overburdened waste stream) (Kim and Rigdon, 1998: 25). Construction and demolition waste presents a potential health risk and makes up 44% of landfill by mass in some Australian states (Apotheker, 1992; CIRIA, 1993; Lahner and Brunner, 1994; McDonald, 1996; 1998; Lingard et al., 2001). For this reason, it is important to reuse or recycle materials wherever possible. Information on these two methods has been briefly provided next (Sertyesilisik et al., 2012):

- Reuse of the materials obtained from demolished buildings can be increased by creating markets for recovered materials and encouraging voluntary co-operation by promoting awareness of disposal problems (Sertyesilisik et al., 2012).
- An important aspect of sustainability is to conserve resources through recycling (Srour et al., 2010) as the World Watch Institute estimated that by 2030 the world will run out of several raw materials for construction (Gorgolewski, 2006). Some environmental and economic constraints to recycling are the energy consumed and pollution created by haulage, transport costs of the material and technical standards and specification of the use of secondary aggregates (Sertyesilisik et al., 2012). Recycling incurs costs for workers in that separating waste for recycling slows the work rate down (Lingard et al., 2001: 814). The six major factors impeding construction materials recycling can be summarised from Srour et al. (2010) as follows: social factors (Teo and Loosemore, 2001); environmental factors (Sheridan et al., 2000); economic factors (Ueda et al., 2003); materials and natural factors; stakeholders' factors (Ruff et al., 1996) and regional factors. These barriers can be overcome by Lingard et al. (2001: 814): rewards for meeting waste management objectives to ensure that workers perceive benefits in waste management; collaboration between main and subcontractors and reassuring subcontractors that their efforts will be rewarded. Furthermore, Srour et al. (2010) quoting from Matten (2010) emphasised the role of strict regulations in encouraging companies to recycle.

4.5 Lean construction

The footprint of the construction industry can be reduced with the help of construction project management enhanced benchmarking management principles in different industries. Accordingly, lean management principles can be adapted to the construction industry, under the name of lean construction, to enhance the

sustainability performance of the construction project management as lean principles lead to a decrease in waste and an increase in value-added activities. Howell
(2001) defines "lean" as giving the customers what they want, when they want,
without waste. According to the lean production principle, everything that is not
contributing to the production process and does not create value is categorised as
"waste". Based on Ohno's categorisation and on the lean principles, waste arises
due to (King, 2009: 37): overproduction, waiting time, transportation, processing itself, inventory, movement and making defective parts. Lean is based on the
following principles: elimination of waste, continuous improvement, just-in-time
production (production on time just when it is needed) and ensuring quality
through suspending the production process until the defects are removed. Lean
leads to an increase in value-added activities and a decrease in waste through con-
tinous improvement (Carroll, 2008: 45). As *leanness* means developing a value
stream to eliminate all waste types (Naylor et al., 1999: 108), *lean construction* can be defined as a continuous improvement to the construction project by
reducing waste of resources, and increasing productivity, leading to enhanced
health and safety as well as sustainability performance (Marhani et al., 2012).
For this reason, lean production leads to the elimination of waste and allows
project managers to "do more with less and less" (Vais et al., 2006). In order to
transform a process into a lean and a green or sustainable one, there is a need
to focus especially on the elimination of wasted materials and energy consumed
(Vais et al., 2006). Lean principles can enhance the sustainability performance
of the construction project management as they support the construction process to become environmentally friendly, energy efficient and waste-free (Arbos,
2002: 169; Abdek-Razek et al., 2007; Marhani et al., 2012: 90) catering for
the client's, communities' and environment's needs of creating value (Marhani
et al., 2012: 91). Adoption of lean principles requires the construction companies
to have an organisation-wide lean process perspective, commitment to become
lean, cultural transformation, as well as education and training (Carroll, 2008:
18–21). Lean construction can be supported by lean tools, which include but are
not limited to (Harris and McCaffer, 1997; Tommelein, 1998; Thomas et al.,
2002; Alinaitwe, 2009; King, 2009; Marhani et al., 2012: 93): just-in-time, total
quality management, business process re-engineering and value based management. Recommended reading on lean tools includes: Koskela (1992); Ballard and
Howell (1994, 1995, 1998); Fisher (1995); Osman and Abdel-Razek (1996);
Harris and McCaffer (1997); Olomolaiye (1998); Tommelein (1998); Thomas
et al. (2002); Bertelsen (2004); Bicheno (2004); Constructing Excellence (2004);
Vais et al. (2006); Alinaitwe (2009); King (2009); Seppanen et al. (2010) and
Marhani et al. (2012). Pre-construction and construction phases are the most
convenient phases for integrating lean construction principles (Marhani et al.,
2012: 95) so they can enable the construction companies to meet the lean targets more easily (Sertyesilisik, 2014). Accordingly, in the planning phase, the
creation of value and the continuous flow of the process need to be considered
(Koskela, 1992; Koskela and Huovila, 1997; Howell, 1999). Controlling is
important throughout project life (Howell, 1999) as it affects the waste generated due to poor workmanship or defects. Controlling can be supported by IT
including geographic information systems and building information modelling
(Irizarry et al., 2013). The construction phase can be supported by lean tools and

principles to eliminate waste and to create value. For example, the construction phase can be enhanced through visual management, which facilitates effective communication among staff, especially through keeping the process areas clean and well organised (King, 2009: 148).

4.6 Conclusion

As the construction industry's footprint affects the environment not only through its output but also through the construction project management process, there is a need to enhance sustainability performance in construction project management. For this reason, this chapter has focused upon the sustainable building process and provided a detailed analysis of sustainable construction in three main phases, namely: pre-construction phase; construction phase and post-construction phase.

Sustainability performance in construction project management can be enhanced through benchmarking management principles and techniques used in other industries. Lean management principles can be adapted to the construction industry and to all phases of construction project management, under the name of lean construction, to reduce waste and to create value.

Sustainable building processes, in the near future, can be expected to be more IT based enabling creation of synergy throughout the construction project management and in the reduction of waste both in terms of lean and sustainability perspectives. Sustainable building processes need to create value not only for today's stakeholders but also for the next generations and for the natural world.

The demand for sustainable buildings and sustainable building processes is expected to increase in the near future especially due to: the increased consciousness of stakeholders with respect to the importance of sustainable buildings and sustainable building process; possible environmental footprint taxes and changes in the accounting system considering the impact on the natural world; as well as due to the EU 20-20-20 targets for reducing CO_2 emissions. For this reason, the motivation of construction companies as well as of construction material production companies to enhance their sustainability performance is expected to increase substantially in the near future.

The next chapter focuses on sustainable buildings themselves. It outlines: the definition of and drivers for sustainable buildings; building assessment tools and key technical aspects of sustainable buildings.

References

Construction contracts and specifications

Arain, F.M., Assaf, S. and Low, S.P. (2004). Causes of Discrepancies between Design and Construction. Architectural Science Review, 47(3), 237–249.

ASHRAE Guideline 1-1996 (1996). The HVAC Commissioning Process, The American Society of Heating, Refrigerating and Air-Conditioning Engineers, Atlanta, US, ISSN: 1049-894X.

Barrett, P.S., Sexton, M.G. and Green, L. (1999). Integrated Delivery Systems for Sustainable Construction. Building Research & Information, 27(6), 397–404.

BRE (2010). The Green Guide to Specifications. (Available at http://www.bre.co.uk, accessed 1 July 2015).

Clough, R.H. and Sears, G.A. (1994). Construction Contracting, 6th ed., Wiley, New York.

Construction Industry Institute (CII) (1986). Impact of Various Construction Contract Types and Clauses on Project Performance, CII, University of Texas at Austin, Austin, TX.

Construction Industry Institute (CII) (1990). The Impact of Changes on Construction Cost and Schedule, CII, University of Texas at Austin, Austin, TX.

Dow, I., Sertyesilisik, B. and Ross, A.D. (2009). An Investigation on the Cost Implications of Methodology of Sub-Contract Work Pricing. Journal of Financial Management of Property and Construction, 14(2), 98–125.

Geok, O.S. (2002). Causes and Improvement for Quality Problems in Design and Build Projects. Unpublished BSc thesis, National University of Singapore, Singapore.

Ibbs, W. and Allen, W.E. (1995). Quantitative Impacts of Project Change, Source Document 108, Construction Industry Institute, University of Texas at Austin, Austin, TX.

Keane, P., Sertyesilisik, B. and Ross, A.D. (2010). Variations and Change Orders on Construction Projects. ASCE Journal of Legal Affairs and Dispute Resolution in Engineering and Construction, 2(2), 89–97.

Koltz, L. and Horman, M. (2010). Counterfactual Analysis of Sustainable Project Delivery Processes. Journal of Construction Engineering and Management, 136(5), 595–605.

Lam, P.T.I., Chan, E.H.W., Chau, C.K., Poon, C.S. and Chun, K.P. (2010a). Environmental Management System vs Green Specifications: How do they complement each other in the construction industry? Journal of Environmental Management. 92(3), 788–795.

Lam, P.T.I., Chan, E.H.W., Poon, C.S., Chau, C.K. and Chun, K.P. (2010b). Factors Affecting the Implementation of Green Specifications in Construction. Journal of Environmental Management, 91(3), 654–661.

Masterman, J.W.E. (2002). Introduction to Building Procurement Systems, 2nd ed., Spon Press, London.

O'Brien, J.J. (1998). Construction Change Orders, McGraw-Hill, New York.

Peace, S. and Bennett, J. (2003a). How to Use a Traditional Approach for a Construction Project: A client guide, Chartered Institute of Building, Ascot. (Available at http://products.ihs.com/cis/Doc.aspx?AuthCode=&DocNum=265786, accessed 30 July 2015).

Peace, S. and Bennett, J. (2003b). How to Use a Design Build Approach for a Construction Project: A client guide, Chartered Institute of Building, Ascot. (Available at http://www.google.com.tr/url?sa=t&rct=j&q=&esrc=s&source=web&cd=5&ved=0CDoQFjAEahUKEwjU-cfFloHHAhVGhiwKHRytBxw&url=http%3A%2F%2Fwww.cvf.or.kr%2Fforum%2Fresearch%2Fpds%2Fdownload.asp%3Fgubun%3Dresearch_pds%26file_name%3DDesign.pdf%26file_name2%3DDesign.pdf&ei=FDe5VdSfFMaMsgGc2p7gAQ&usg=AFQjCNFWwt7ddIo6jn5TfSTOHpHzWfU0hQ, accessed 30 July 2015).

Sertyesilisik, B. (2010). Investigation on Particular Contractual Issues in Construction. Journal of Legal Affairs and Dispute Resolution in Engineering and Construction, 2(4), 218–227.

Construction project management

Abdel-Razek R., Elshakour H.A. and Abdel-Hamid, M. (2007). Labour productivity: benchmarking and variability in Egyptian projects. International Journal of Project Management. 25(2), 189–197.

Construction Industry Institute (CII) (1994). Pre-Project Planning: Beginning a Project the Right Way, Publication 39-1, CII, University of Texas at Austin, Austin, TX.

Dell'Isola, A.J. (1982). Value Engineering in the Construction Industry, 3rd ed., Van Nostrand Reinhold, New York.

Fisk, E.R. (1997). Construction Project Administration, 5th ed., Prentice-Hall, Upper Saddle River, NJ.

Glavinich, T.E. (2008). Contractor's Guide to Green Building Construction: Management, Project Delivery, Documentation and Risk Reduction, John Wiley & Sons, Inc., Hoboken, NJ.

Holtom, B.C., Mitchell, T.R., Lee, T.W. and Inderrieden, E.J. (2005). Shocks as Causes or Turnover: What they are and how organisations can manage them. Human Resource Management, 44(3), 337–352.

Jones, S.M., Ross, A. and Sertyesilisik, B. (2010). Testing the Unfolding Model of Voluntary Turnover on Construction Professionals. Construction Management and Economics, 28(3), 271–285.

Ofori, G. (1992) The Fourth Construction Project Objective? Construction Management and Economics, 10, 369–395.

Wang, Y. (2000). Coordination Issues in Chinese Large Building Projects. Journal of Management in Engineering, 16(6), 54–61.

Yudelson, J. (2007). Green building A to Z: Understanding the Language of Green Building, New Society Publishers, Canada.

Deep learning

Al-Khaddar R., Wooder, T., Sertyesilisik, B. and Tunstall, A. (2012). Deep Learning Approach's Effectiveness on Sustainability Improvement in the UK Construction Industry. Management of Environmental Quality: An International Journal, 23(2), 126–139.

Ramsden, P., Beswick, D. and Bowden, J. (1989). Effects of Learning Skills Intervention on First Year Students' Learning. Human Learning, 5(3), 151–164.

Warburton, K. (2003). Deep Learning and Sustainability. International Journal of Sustainability in Higher Education, 4(1), 44–56.

Design

Cowan, S. and van der Ryn, S. (1996) Ecological Design, Island Press, Washington, DC.

Good, N. Eco-Charrette. (Available at http://www.betterbricks.com/DetailPage.aspx?Id=275, accessed 1 July 2015).

ICAEN (InstitutCatalad'Energia). (2004). Sustainable Building Design Manual. The Energy and Resources Institute, New Delhi, India.

Kim J.J. and Rigdon B. (1998). Sustainable Architecture Module: Introduction to Sustainable Design. National Pollution Prevention Center for Higher Education. (Available at www.css.snre.umich.edu/~nppcpub/, accessed 1 July 2015).

Integrated design

Busby, Perkins and Will. (2007). Roadmap for the Integrated Design Process. Stantec Consulting.

California Energy Commission. (2007). (Available at http://www.energy.ca.gov/, accessed 30 July 2015).

MFE Website. Integrated Whole Building Design Guidelines. (Available at http://www.mfe.govt.nz/sites/default/files/integrated-building-guidelines.pdf, accessed 30 July 2015).

Reed, W. and Gordon, E. (2000). Integrated Design and Building Process: What Research and Methodologies are Needed? Building Research & Information, 28(5), 325–337.

WBDG Aesthetics Subcommittee. (2009). Whole Building Design Guide Engage the Integrated Design Process. (Available at http://www.wbdg.org/, accessed 30 July 2015).

Lean construction

Alinaitwe, H.M. (2009). Prioritising Lean Construction Barriers in Uganda's Construction Industry. Journal of Construction in Developing Countries, 14(1), 15–30.

Arbos, L.C. (2002). Design of a rapid response and high efficiency service by lean production principles: Methodology and evaluation of variability of performance. International Journal of Production Economics 80(2), 169–183.

Ballard, G. and Howell, G. (1994). Implementing Lean Construction: Improving Downstream Performance. In: Proceedings of the Second Annual Conference of the International Group for Lean Construction, Santiago, Chile.

Ballard, G. and Howell, G. (1995). Implementing Lean Construction: Stabilizing Work Flow. In: Proceedings of the Second Annual Conference of the International Group for Lean Construction, Santiago, Chile.

Ballard, G. and Howell, G.A. (1998). Shielding Production: An Essential Step in Production Control. Journal Construction Engineering and Management, ASCE, Reston, VA. (Available at http://www.cce.ufl.edu, accessed 1 July 2015).

Bertelsen, S. (2004). Lean Construction: Where are we and how to proceed. (Available at http://www.kth.se, accessed 1 July 2015).

Bicheno, J. (2004). The New Lean Toolbox Towards Fast, Flexible Flow. Picsie Books, Buckingham, England.

Carroll, B.J. (2008). Lean Performance ERP Project Management: Implementing the Virtual Lean Enterprise, 2nd ed., Auerbach Publications Taylor & Francis Group.

Constructing Excellence. (2004). Effective Teamwork: A Best Practice Guide for the Construction Industry, Constructing Excellence, London, pp. 1–20.

Fisher, D. (1995). Benchmarking in Construction Industry. Journal of Management in Engineering, 11(1), 50–57.

Harris, F. and McCaffer, R. (1997). Modern Construction Management, Blackwell Science, London.

Howell, G. (2001). Introducing Lean Construction: Reforming Project Management. Report Presented to the Construction User Round Table (CURT), Lean Construction Institute.

Howell, G.A. (1999). What is Lean Construction? IGLC-7 Proceedings, 26–28 July 1999, University of California, Berkeley, CA.

Irizarry, J., Karan, E.P. and Jalaei, F. (2013). Integrating BIM and GIS to Improve the Visual Monitoring of Construction Supply Chain Management. Automation in Construction, 31, 241–254.

King, P.L. (2009). Lean for the Process Industries Dealing with Complexity, CRC Press, Taylor & Francis Group, A Productivity Press Book.

Koskela, L. (1992). Application of the New Production Philosophy to Construction, Technical Report No. 72, CIFE, Stanford University, CA.

Koskela, L. and Huovila, P. (1997). On Foundations of Concurrent Engineering. Proc. Concurrent Engineering in Construction CEC'97. In: Anumba C, Evbuomwan N (eds.). Paper presented at the 1st International Conference on Concurrent Engineering in Construction, London, 3–4 July. The Institution of Structural Engineers, London, 22–32.

Lapinski, A., Horman, M. and Riley, D. (2006). Lean Processes for Sustainable Project Delivery. Journal of Construction Engineering and Management, 132(10), 1083–1091.

Marhani, M.A., Jaapar, A. and Bari, N.A.A. (2012) Lean Construction: Towards Enhancing Sustainable Construction in Malaysia. Procedia – Social and Behavioral Sciences, 68, 87–98.

Naylor, J.B., Naim, M.M. and Berry, D. (1999). Leagility: integrating the lean and agile manufacturing paradigms in the total supply chain. International Journal of Production Economics, 62(1–2), 107–118.

Olomolaiye, P.O. (1998). Construction Productivity Management. Addison Wesley Longman Limited, Edinburgh Gate, England.

Osman, I. and Abdel-Razek, R.H. (1996). Measuring for Competitiveness: The Role of Benchmarking. In: Proceedings of the Cairo First İnternational Conference on Concrete Structures, Cairo University, Cairo, 2–4 January 1996, Vol. 1, 5–12.

Seppanen, O., Ballard, G. and Pesonen, S. (2010). The Combination of Last Planner System and Location Based Management System. (Available at http://www.lean.org, accessed 1 July 2015).

Sertyesilisik, B. (2014). Lean and Agile Construction Project Management: As a Way of Reducing Environmental Footprint of the Construction İndustry. In: Xu, H. and Wang, X. (eds.), Optimisation and Control Methods in Industrial Engineering and Construction Intelligent Systems, Control and Automation: Science and Engineering, Vol. 72, 179–196.

Thomas, H.R., Michael, J.H. and Zavrski, I. (2002). Reducing Variability to Improve Performance as a Lean Construction Principle. Journal of Construction Engineering and Management,128(2), 144–154.

Tommelein, I.D. (1998). Pull-Driven Scheduling for Pipe Spool Installation: Simulation of Lean Construction Technique. Journal of Construction Engineering and Management, 124(4), 279–288.

Vais, A., Miron, V., Pedersen, M. and Folke, J. (2006). "Lean and Green" at a Romanian Secondary Tissue Paper and Board Mill – Putting Theory Into Practice. Resources, Conservation and Recycling, 46, 44–74.

Materials

Kim, J.-J. and Rigdon, B. (1998). Qualities, Use and Examples of Sustainable Building Materials. National Pollution Prevention Center for Higher Education, Ann Arbor, MI. (Available at http://www.umich.edu/~nppcpub/resources/compendia/architecture.html#ranr, accessed 1 July 2015).

Lahner, T.E. and Brunner, P.H. (1994). Buildings as Reservoirs of Materials: Their Reuse and Implications For Future Design. In: E.K. Lauritzen (ed.), Demolition and Reuse of Concrete, E. & F.N. Spon, London.

Morel, J.C., Mesbah, A., Oggero, M. and Walker, P. (2001). Building Houses with Local Materials: Means to Drastically Reduce the Environmental Impact of Construction. Building and Environment, 36(10), 1119–1126.

Sustainable construction

Architectural Research Centers Consortium. (2007, Spring). Conference, University of Oregon, School of Architecture and Allied Arts, April 16–18, 2007, 191–192.

ASHRAE. (2006). The ASHRAE Green Guide, 2nd ed., ASHRAE Press, 3–100.

Burnett, J. (2007). City building- Eco-labels and shades of green. Landscape and urban planning, 83., 29–38.

CIRIA. (1993). Environmental Issues in Construction: A Review of Issues and Initiatives Relevant to the Building, Construction and Related Industries, Vol. 2, Construction Industry Research and Information Association, London.

Cole, R.J. (1999). Building environmental assessment methods: clarifying intentions. Building Research and Information, 27(4/5), 230–246.

Ding, G.K.C. (2005). Developing a Multicriteria Approach for the Measurement of Sustainable Performance. Building Research and Information, 33(1), 3–16.

Djuric, N. and Novakovic, V. (2010). Correlation between Standards and the Lifetime Commissioning. Energy and Buildings, 42, 510–521.

Drager, L. (1996). An Investigation into the State of Sustainable Construction within the South African Building Industry. Report, Department of Environmental and Geographical Science, University of Cape Town, South Africa.

Du Plessis, C., Adebayo, A., Agopyan, V., Beyers, C., Chambuya, S., Ebohon, J., Giyamah, O., Irurah, D., John, V., Hassan, A., Laul, A., Marulanda, L., Napier, M., Ofori, G., Pinto de Arruda, M., Rwelamila, P.D., Sara, L., Sattler, M., Shafii, F., Shah, K., Sara, L. and Sjostrom, C. (2002). Agenda 21 for Sustainable Construction in Developing Countries: A Discussion Document, Johannesburg: Commissioned for the World Summit on Sustainable Development. Report No. Bou/EO 204, CSIR Building and Construction Technology (Boutek), Pretoria.

Hill, R.C. and Bowen, P.A. (1997). Sustainable Construction: Principles and a framework for attainment. Construction Management and Economics, 15, 223–239.

Holmes, J. and Hudson, G. (2000). An Evaluation of the Objectives of the BREEAM Scheme for Offices: A Local Case Study. In: Proceedings of Cutting Edge 2000, RICS Research Foundation, RICS, London.

Kibert, C.J. (2007). The Next Generation of Sustainable Construction. Building Research and Information, 35(6), 595–601.

Masstech Website. (Available at http://www.masstech.org/, accessed 30 July 2015).

Ross, N., Anthony, B.P. and Lincoln, D. (2010). Sustainable Housing for Low-Income Communities: Lessons for South Africa in local and other developing world cases. Construction Management and Economics, 28(5), 433–449.

Scheuer, C., Keoleian, G.A. and Reppe, P. (2003). Life Cycle Energy and Environmental Performance of a New University Building: Modelling Challenges and Design Implications. Energy and Buildings, 35, 1049–1064.

Shen, L. and Tam, V. (2002). Implementation of Environmental Management in the Hong Kong Construction Industry. International Journal of Project Management, 20(7), 535–543.

Recycling and reuse

Gorgolewski, M. (2006). The Implications of Reuse and Recycling for the Design of Steel Buildings. Canadian Journal of Civil Engineering, 33, 489–496.

Lund, H.F. (2001). The McGraw-Hill Recycling Handbook, Second edition. McGraw-Hill: New York.

Merry, W. (1990). Taking a Pro Table Approach To Recycling. World Wastes, July, 40–47.

Srour, I., Chong, W.K. and Zhang, F. (2010). Sustainable Recycling Approach: An Understanding of Designers' and Contractors' Recycling Responsibilities Throughout the Life Cycle of Buildings in Two US Cities. Sustainable Development, 20(5), 350-360. (Available at http://onlinelibrary.wiley.com/doi/10.1002/sd.493/pdf, accessed 30 July 2015).

Ueda, K., Nishino, N. and Oda, S.H. (2003). Integration of economics into engineering with an application to the recycling market. CIRP Annals–Manufacturing Technology, 52(1), 33–36.

Waste management

Apotheker, S. (1992). Managing Construction and Demolition Materials. Resource Recycling, August, 50–61.

Baetz, B., Pas, E. and Vesilind, P. (1991). Waste-Reduction Primer for Managers. Journal of Management in Engineering, 7(1), 33–42.

Beaumont, J.R., Pedersen, L.M. and Whitaker, B.D. (1993). Managing the Environment, Butterworth Heinemann, Oxford.

Chung, S.S. and Lo, C.W.H. (2003). Evaluating Sustainability in Waste Management: The case of construction and demolition, chemical and clinical wastes in Hong Kong. Resources, Conservation and Recycling, 37, 119–145.

Cosper, S.D., Hallenbeck, W.H. and Brenniman, G.R. (1993). Construction and Demolition Waste: Generation, Regulation, Practices, Processing and Policies, Office of Solid Waste Management, The University of Illinois, Chicago.

Federle, M.O. (1993). Overview of Building Construction Waste and the Potential for Materials Recycling. Building Research Journal, 2, 31–37.

Hao, J., Hill, M. and Shen, L. (2007). Managing Construction Waste On-Site through System Dynamics Modelling. Engineering, Construction and Architectural Management, 15(2), 103–113.

Johnston, H. and Mincks, W. (1995). Cost-Effective Waste Minimization for Construction Managers. Journal of Cost Engineering, 37(1).

Lingard, H., Gilbert, G. and Graham, P. (2001). Improving Solid Waste Reduction and Recycling Performance using Goal Setting and Feedback. Construction Management and Economics, 19(8), 809–817.

Lingard, H., Graham, P. and Smithers, G. (2000). Employee Perceptions of the Solid Waste Management System Operating in a large Australian Contracting organization: implications for company policy implementation. Construction Management and Economics, 18(4), 383–393.

Matten, D. (2010). Enforcing Sustainable Development by Legislation: Entrepreneurial Consequences of the New German Waste Management Act. Sustainable Development, 4(3), 130–137.

McDonald, B. (1996). RECON Waste minimisation and environmental programme, CIB TG16 Commission meetings and presentations, Melbourne.

McDonald, B. and Smithers, M. (1998). Implementing a Waste Management Plan during the Construction Phase of a Project: A case study. Construction Management and Economics, 16, 71–8.

Ortiz, O., Bonnet, C., Bruno, J.C. and Castells, F. (2009). Sustainability Based on LCM of Residential Dwellings: A case study in Catalonia, Spain. Building and Environment, 44, 584–594.

Osmani, M., Glass, J. and Price, A.D.F. (2008). Architects' Perspectives on Construction Waste Reduction by Design. Waste Management, 28, 1147–1158.

Poon, C., Yu, T. and Ng, L. (2001). On-Site Sorting of Construction and Demolition Waste. Resources, Conservation and Recycling, 32(2), 157–172.

Ruff, C.M., Dzombak, D.A. and Hendrickson, C.T. (1996). Owner–contractor Relationships on Contaminated Site Remediation Projects. Journal of Construction Engineering and Management, 122(4), 348–353.

Schlauder, R.M. and Brickner, R.H. (1993). Setting Up for Recovery of Construction and Demolition Waste. Solid Waste and Power, January/February, 28–34.

Sertyesilisik, B., Remiszewski, B. and Al-Khaddar, R. (2012). Sustainable Waste Management in the UK Construction Industry, 4(2), 173–188.

Sheridan, S., Townsend, T., Price, J., and Connell, J. (2000). Policy options for hazardous-building-component removal before demolition. Practice Periodical of Hazardous, Toxic, and Radioactive Waste Management, 4(3), 111–117.

Teo, M.M. and Loosemore, M. (2001). A Theory of Waste Behaviour in the Construction Industry. Construction Management and Economics, 19(7), 741–751.

Von Stein, E.L. and Savage, G.M. (1994). Current Practices and Applications in Construction and Demolition Debris Recycling. Resource Recycling, April, 85–94.

5   Sustainable Buildings

Begum Sertyesilisik

This chapter on sustainable buildings focuses on: the definition of and drivers for sustainable buildings, building assessment tools and key technical aspects of green and sustainable buildings.

5.1 Definition of and drivers for sustainable buildings

Natural resources are being depleted at a rate faster than their replenishment (Lam et al., 2010). Buildings, their surroundings and related enterprises produce more CO_2, generate more pollution, consume more energy and waste more natural resources than any other human enterprise or industry (Sozer, 2010: 2581). They are responsible for 38% of all CO_2 emissions (EIA, 2008, as quoted by Tatari and Kucukvar, 2010: 2) and the building industry consumes one-half of the world's physical resources (RCA website as quoted by Lam et al., 2010). All of these factors mean that demand for sustainable construction is increasing.

Kibert (1994) defined "sustainable construction" as the creation and responsible management of a healthy built environment based on resource efficient and ecological principles. Sustainable construction is how the construction industry, together with its product the "built environment", can contribute to the sustainability of the earth (Kibert, 2007a: 595). Sustainable construction (Du Plessis, 2007: 69–70):

- emphasises environmental protection and value addition to individuals' and communities' quality of life
- embraces technological responses as well as the non-technical aspects related to social and economic sustainability.

Sustainable construction consists of four attributes (Hill and Bowen, 1997), namely: social, economic, biophysical and technical attributes establishing a framework for achieving sustainable development that includes an environmental assessment during the planning and design stages of projects and the implementation of environmental management systems (Ding, 2005: 5). There are 10 "action points" to achieve more sustainable construction (Vatalis et al., 2011: 378): reuse of built assets, design for minimum waste, the aim of lean construction, minimising energy in construction, minimising energy in use, restraint in pollution, preserving and enhancing biodiversity, conservation of water resources, respect for people and their local environment and target setting (DETR, 2000).

A green building is designed to use less energy and water and to reduce the life-cycle environmental impacts of the materials used (Yudelson, 2008: 13). According to Kibert and Grosskopf (2005), the main characteristics of green buildings are: closed loop material systems, integration with local ecosystems, maximum use of passive design and renewable energy, full implementation of indoor environmental quality measures and optimised building hydrologic cycles (Burnett, 2007: 30). Closed loop material systems enable materials to be reused or recycled. Silvestre et al. (2014) states that "closing material loops can be achieved either by designing buildings for deconstruction or from developing building products that can be dismantled; both options are being increasingly addressed in the context of green buildings (IEA, 2004; Kibert, 2007a)". Passive design reduces the need for artificial energy and enhances the sustainability performance of the construction. Passive design enables the building to benefit from the natural environment (i.e. orientation of the building towards or away from the sun depending on the climatic conditions; considering the wind velocity via the distances among buildings, natural ventilation that can be carried out due to the thermal forces created by differences between indoor and outdoor temperatures). As passive design focuses on maintaining energy with passive methods including insulation, an "air-tight envelope", or thermal material (Whang and Kim, 2014: 304), passive design strategies contribute to improving the interior comfort conditions, increasing the energy efficiency in buildings and reducing their energy consumption (Rodriguez-Ubinas et al., 2014).

Today's high-performance green buildings are a significant improvement over the conventional buildings of the past as they consume significantly less energy, fewer materials and less water; they provide healthy living and working environments and greatly improve the quality of the built environment (Kibert, 2007a: 595). Contractors therefore need to make the environment a key element in their business strategy and day-to-day operations and to investigate ways to promote and demonstrate commitment to the environment (Glavinich, 2008: 4–6). Contractors can be encouraged by the drivers of sustainable construction (Table 5.1).

Despite these drivers, contractors face obstacles that can discourage them from carrying out sustainable construction. Issues contractors have to consider (CIB, 1999) (as quoted from Huang and Hsu, 2011: 150) can be categorised into: process and management related issues (e.g. design process, decision-making processes); product and building related issues (e.g. indoor environmental quality, reparability); resource consumption related issues (e.g. energy efficiency, design for a long service life); built environment sustainability related issues (e.g. environment quality, life quality); social issues (e.g. creating a safe and healthy

Table 5.1 Drivers for sustainable construction

• Desire for sustainable development in global terms	Doughty and Hammond (2004: 1223)
• Trend for taking key business decisions with environmental, social and economic concerns • The economic benefits of sustainable construction • A healthy comfortable living environment • Reduction in energy usage • The natural and social environment	Vatalis et al. (2011)

Table 5.1 Drivers for sustainable construction (*cont.*)

• Demonstration by contractors and their suppliers that they are committed to the environment and provide environmentally sustainable products and services • Employees' motivation and willingness to work for an environmentally-conscious firm • Improved productivity • Reduced costs at the jobsite • Reduced home office overhead costs • Increase in the effectiveness of construction team members • Appeal to many of the contractor's clients • Benefits for the contractor, its employees, other stakeholders and society	Glavinich (2008: 3–6)
• Contribution to public health and the environment • Reduction in operating costs • Enhancing building and organisational marketability • Increase in occupant productivity • Creation of a sustainable community	Green Building Rating System For Commercial Interiors Version 2.0 Updated December 2005

working environment) and economic issues (e.g. acquiring financial benefits and uplift for the community).

Sustainable construction faces economic challenges at three levels: macro, meso and micro (Vatalis et al., 2011: 378–379). At the macro level the challenge is particularly acute in developing countries, where the demand for construction is growing. As a result, achieving the required levels of construction is seen as more important than building sustainably. At the meso level the challenge lies in assessing the impact of different materials and processes across the supply chain, which is long and complex. At the micro level, the challenge is getting people to think about the long-term financial benefits of sustainable construction, rather than focusing on the short and medium term (Bon and Hutchinson, 2000).

Cost-benefit analysis is important for understanding the project's feasibility. The cost-benefit analysis method aims to analyse whether the total benefits of a project exceed the total costs (Ding, 2005: 5). This method is, however, criticised for ignoring the value of environmental goods and services (Tisdell, 1993; Hobbs and Meier, 2000; RICS, 2001; Ding, 2005: 5). Project costs and benefits should not only be calculated based on market transaction and price, but also on the monetary value of the estimated assets (Ding, 2005: 5). Vatalis et al. (2011: 378) described the difficulty in calculating the economic benefits of sustainable buildings due to the uniqueness of each project. Each has its own financial needs and performance data, which makes it difficult to generalise when discussing sustainable solutions throughout the life cycle.

5.2 Building assessment tools

Building assessment tools (BATs) have been developed as voluntary instruments to provide a catalyst for market transformation (Cole, 2003). The requirements of these certificates impact building design and construction (Glavinich, 2008: 15). BATs first emerged in the 1990s (Burnett, 2007: 31) and there are more than 70 tools for evaluating and classifying building projects based on sustainability indicator systems (Fernández Sánchez, 2008) including: BREEAM (Baldwin et al., 1990, 1998), HK-BEAM (CET, 1996; HK-BEAMSociety, 2004), LEED (US Green Building Council, 1999, 2003), CASBEE (Institute of Building Environment and Energy Conservation, 2003) and Green Star (Green Building Council of Australia, 2005). Performance assessment is at the core of sustainable construction (Huang and Hsu, 2011: 144). The outcome of the assessment is an eco-label based on the sum of points or credits obtained (Burnett, 2007: 31).

BATs have four aims. First they evaluate the environmental characteristics of the buildings by using a set of standards for the achievement of more environmentally friendly building performance (Cole, 1998; Tatari and Kucukvar, 2010: 4). Second, they stimulate the market demand for buildings with improved environmental performance by providing consumers with an extra reference for making rental or purchase decisions (Crawley and Aho, 1999). Next, they fulfil the sustainability goal of the assessment method for developers and they minimise the overall building impact on the environment (Chau et al., 2000: 963). Finally they promote building systems' integration and the optimisation of the building as a whole (Glavinich, 2008: 17).

The BATs face criticism on a number of levels:

- As sustainable building includes environmental, economic and social aspects, transforming the existing building environmental assessment tools into sustainability assessment tools seems, at the moment, far away (Haapio and Viitaniemi, 2008: 480).
- These methods mainly focus on the assessment of sustainability performance at the project level and fail to evaluate the overall performance of sustainable construction from a national viewpoint, which is important to establish common targets toward the industry's sustainability (Huang and Hsu, 2011: 145).
- "Most of the existing assessment schemes do not provide sufficient guidance to help the investors prioritise the environmental improvement measures for selection, since they predominately focus on reducing negative environmental impacts (Finch, 1992; Levin, 1997; Cole, 1998)." (Chau et al., 2000: 961).
- Although most of the existing schemes are voluntary, many fail to encourage participation of the building investors and designers (Chau et al., 2000: 259).
- "There is no consensus-based approach to guide the assignment of weightings (Bisset, 1980; Levin, 1997; Cole, 1998; Wehrmeyer and Tyteca, 1998) affecting the final outcome of the buildings' performance" (Chau et al., 2010: 961).
- BATs present problems such as (Fernández-Sánchez and Rodríguez-López, 2010: 1193): "uncertainty and subjectivity when selecting criteria, indicators

and dimensions (Hueting and Reijnders, 2004; Seo et al., 2004); the pre-domination of environmental aspects when evaluating the sustainability of buildings (Saparauskas, 2007); the lack of participation of all stakeholders involved in the project life cycle (Fernández Sánchez, 2008) and the number of indicators that generally should be small and in the existing systems of indicators is very high (Alarcón Núñez, 2005)."

BATs are similar in: the green building criteria that they address; their usage as guidance to the design and construction of green buildings (Glavinich, 2008: 16) and their intent to enhance the environmental performance of buildings, by identifying significant environmental impacts promoting positive impacts over a building's life cycle (Burnett, 2007: 36). BATs have to be tailored for the local context and are therefore usually specific to one country. They include criteria for the most significant external environmental impacts, global and regional impacts (such as GHG emissions, ozone depletion, deforestation, NO_x, SO_x, particulate emissions, river pollution etc.), local impacts (including waste, water and local air pollution) and neighbourhood impacts (such as overshadowing, noise from building equipment etc.) (Burnett, 2007: 34).

BATs can be different in their intent, criteria, emphasis and implementation, especially due to the sponsoring organisation's goals (Glavinich, 2008: 16). They also vary in: building types assessed, phases emphasised, environmental issues considered (i.e. global, national and local), their purposes (i.e. research, consulting, decision making and maintenance), their users (i.e. designers, architects, researchers, consultants, owners, tenants and authorities) (Haapio and Viitaniemi, 2008: 469–470). Furthermore, they vary in their performance requirements (criteria, levels, standards etc.), assessment methods demonstrating compliance and the scoring system that determines the final grade (eco-label) (Burnett, 2007: 31).

The following sections take a closer look at the most commonly used green building certificates, which are LEED (the Leadership in Energy and Environmental Design), BREEAM (Building Research Establishment's Environmental Assessment Method), CASBEE (Comprehensive Assessment System for Building Environmental Efficiency) and Green Star.

5.2.1 LEED

USGBC (U.S. Green Building Council) is an industry organisation promoting the construction of environmentally friendly buildings through its sponsorship of the Leadership in Energy and Environmental Design (LEED™) green building rating systems (Glavinich, 2008: 17–18). The USGBC guiding principles include (but are not limited to) the promotion of the triple bottom line solutions, which promote harmony between social, environmental and economic aspects and demonstrating transparency in the building process (Montoya, 2011: 106).

The LEED™ rating system is the most widely used assessment tool (Tatari and Kucukvar, 2010: 4). LEED™ certification requirements need to be followed in the design process, and implemented in the construction process (Glavinich, 2008: 18). LEED™ has the following rating systems: New Construction; Existing Buildings: Operations & Maintenance; Commercial Interiors; Core and Shell;

Schools; Retail; Healthcare; Homes and Neighborhood Development. LEED™ was created primarily to transform the building market into a more sustainable one, particularly through encouraging green competition, increasing awareness among consumers and creating a standard of measurement for green building (Montoya, 2011: 110).

LEED™ Rating Systems can be useful as they provide:

- recognition for a construction company's commitment to environmental issues in its community, its organisation (including stockholders) and its industry;
- third party validation of achievement;
- qualification for a growing array of state and local government initiatives;
- "marketing exposure through USGBC website, Greenbuild conference, case studies and media announcements." (USGBC website)

The contractor needs to be aware of the LEED™ requirements, as they can influence material and equipment procurement, construction requirements and costs, allowing the contractor to analyse the project with respect to the LEED™ requirements (Glavinich, 2008: 18). The USGBC website provides detailed information on the LEED™ rating systems and procedures.

The LEED™ certification process starts with online registration where the project goals are specified under six categories, namely: sustainable sites, water efficiency, energy and atmosphere, materials and resources, indoor environmental quality, innovation and design process. The necessary project documentation should be carried out so that the project can show whether or not it complies with all prerequisites needed and the number of credits earned. Assessments are completed either online, or as hard copy (Saunders, 2008: 23). Although it is not compulsory, if a LEED™ accredited professional is appointed within the design team, the project gets a credit for it. Appointing a LEED™ AP Accredited Professional can support the project team with regard to the LEED™ requirements and procedures (Glavinich, 2008). The LEED™ certification level (certified, silver, gold or platinum) is determined based on the number of credits earned. The design team can appeal within 25 days following USGBC's assessment if they are not satisfied with the results.

5.2.2 BREEAM

BREEAM (Building Research Establishment's Environmental Assessment Method) was first launched in 1990 in the UK as a voluntary green building rating system (Saunders, 2008: 32). The BREEAM's scheme manuals are available for different building types, namely (BRE Global, 2009: 13): courts, education, industrial, healthcare, offices, retail, prisons and multi-residential. BREEAM credits are awarded in ten categories, namely: management, health and wellbeing, energy, transport, water, materials, waste, land use and ecology, pollution and innovation (BRE Global, 2009: 14). There are different assessment tools for different stages of the building's life, namely (Saunders, 2008: 32): BREEAM design and procurement; the post-construction review; the fit out assessment and a management and operation assessment.

The BREEAM scheme can be used to assess environmental impacts of an individual building development at the following two stages (BRE Global, 2009: 19):

1. Design Stage leading to an Interim BREEAM Certificate
2. Post-Construction Stage leading to a final BREEAM Certificate after practical completion of the building works.

In case a formal interim design stage assessment has not been carried out and in case BREEAM assessment and rating is required, a full post-construction stage assessment can be conducted as well (BRE Global, 2009: 19). The BREEAM rating depends on (BRE Global, 2009: 25): BREEAM rating benchmarks, BREEAM environmental weightings, minimum BREEAM standards and BREEAM credits for Innovation. The certification level (pass, good, very good, excellent and outstanding) is determined based on the total credits earned (BRE Global, 2009: 8–14). BREEAM assessments are carried out by licensed assessors (organisations or individuals) trained, examined and licensed by the BRE to help design teams (or facilities management companies) to carry out the assessments (Saunders, 2008: 33). The BREEAM third party verification process can be summarised in the following successive steps (Saunders, 2008: 15): registration; issuing of an assessment reference number; information collection by the assessor to check compliance with BREEAM criteria; assessment and rating by the independent BREEAM assessor; submission of the assessor's report; quality assurance check and certification. Case studies for BREEAM certified buildings can be accessed from the following website: http://www.breeam.org/case-studies.jsp.

5.2.3 CASBEE

CASBEE (Comprehensive Assessment System for Building Environmental Efficiency) is a building quality assessment tool, covering aspects such as interior comfort and consideration of scenery, as well as evaluating environmental aspects such as using materials and equipment that save energy or cause smaller environmental loads. CASBEE is a "self-assessment check system", which can also be used as a labelling system, in case the assessment is verified by a third party (Saunders, 2008: 17). CASBEE is mainly used in Japan. The first assessment tool, CASBEE for Office, was completed in 2002, followed by CASBEE for New Construction in July 2003, CASBEE for Existing Building in July 2004 and CASBEE for Renovation in July 2005 (CASBEE website; CASBEE, 2008a, and CASBEE, 2008b). The following CASBEE tools have been developed (CASBEE website): CASBEE for Building (New Construction); CASBEE for Building (Existing Building); CASBEE for Building (Renovation); CASBEE for Market Promotion; CASBEE for Heat Island; CASBEE for Urban Development; CASBEE for Cities; and CASBEE for Home (Detached House). The 2014 edition of CASBEE tools have been released (CASBEE website).

CASBEE assessment tools are based on the following three principles (CASBEE website, CASBEE, 2008a, and CASBEE, 2008b):

- assessment can continue throughout the life cycle of the building
- assessment can consider both the "Environmental quality of the building (Q)" and the "Environmental load of the building (L)"

- the idea of environmental efficiency is employed to evaluate on the basis of Building Environmental Efficiency (BEE), which is the equation of Q/L.

CASBEE uses weightings that are applied to four main categories ("indoor environment", "outdoor environment onsite", "energy" and "resources & materials") each of which covers various headline issues (e.g. "Service Ability", "lighting and illumination" and "building thermal load") to which further weightings are added (Saunders, 2008: 17). There are five ratings: C (poor), B– (fairly poor), B+ (good), A (very good) and S (excellent).

CASBEE comprises four basic tools that correspond to building life-cycle stages, namely: pre-design, new construction, existing buildings and renovation (CASBEE website; CASBEE, 2008a, and CASBEE, 2008b). CASBEE's third-party verification process includes the following steps (Saunders, 2008: 12): report by the CASBEE assessor, acceptance by CASBEE, credit filter check, final credit appraisal, agreement on the precise characteristics of development, submission of the report and quality assurance check and publication of the result. CASBEE certified buildings can be accessed from the following website: http://www.ibec.or.jp/jsbd/.

5.2.4 Green Star

The Green Building Council of Australia (GBCA) is a non-profit organisation committed to developing a sustainable property industry for Australia by encouraging the adoption of green building practices (GBCA, 2008: 1). The GBCA has developed various rating tools for different phases of the building life cycle (design, fit out and operation) and for different building classes including, but not limited to, office, retail, industrial and residential (GBCA, 2005: 1). Green Star's first version was developed in 2003. As BREEAM was used as the basis of the Green Star methodology, adaptations have been carried out due to the differences between Australia and the UK, considering factors such as the climate, local environment and the construction industry standard practice (Saunders, 2008: 27). The potential environmental performance of buildings is evaluated based on a number of criteria including energy and water efficiency, quality of indoor environments and resource conservation (GBCA, 2005). Following the assessment of the credits for each category, a percentage score is calculated and Green Star environmental weighting factors are applied (Saunders, 2008: 29). The following Green Star certified ratings are available: 4-star (best practice), 5-star (Australian excellence) and 6-star (world leadership).

If the final score of the project is below 45 and the project does not qualify for a certified rating, the GBCA does not disclose the results of the Green Star assessment or the project's initial intent of achieving a Green Star Certified Rating until advised otherwise (GBCA, 2009a: 6–7).

Green Star Certification is based on the submittals providing evidence of the sustainability performance achievement of the building (GBCA, 2009a: 1). A panel of third-party Certified Assessors is assigned by the GBCA for the assessment of the project with respect to the Green Star certification requirements (GBCA, 2009a: 1). The assessment is carried out in two rounds (GBCA, 2009a: 1). Project teams are notified of the final score and a framed certificate, award letter, marketing kit as well as relevant Green Star logos are provided to the projects that are awarded a certified rating (GBCA, 2009a: 1). A critical point in the assessment process is that it is the responsibility of the project team to ensure that the project complies with all eligibility criteria at the time of registration as a project that fails to meet all eligibility criteria is not assessed (GBCA, 2009a: 2). Eligibility criteria that need to be fulfilled are (GBCA, 2009b: 2–7): space use, spatial differentiation, conditional requirements and timing of certification. If a project does not achieve its desired Green Star Certified Rating after the Round 2 Assessment, it has the option of lodging a formal Appeal (GBCA, 2009a: 7).

Having so many different BATs poses problems, particularly in an increasingly globalised world. Saunders (2008: 45) proposed a multi-labelling system to help overcome this problem as such a system would allow multinational organisations with buildings in different countries to compare and assess the environmental performance of buildings all over the world. Burnett (2007: 33), on the other hand, emphasised the difficulty of satisfying every individual, stating the following:

> Criteria and standards that define thermal comfort vary between countries, but whichever the criteria chosen and delivered to a space, not all building occupants will be satisfied…because of individual preference.

5.3 Key technical aspects of sustainable buildings

In accordance with the BATs requirements, basic elements of designing a green development include the following (Wilson et al., 1998: 160–163): design for fitting the site, fostering community, adaptability, a healthy indoor environment, resource efficiency and durability. Furthermore, sustainable buildings are designed to have a high thermal efficiency, especially through a careful design of the building envelope and a limited use of active air-conditioning systems during the building's lifespan (Pulselli et al., 2009: 920).

5.3.1 Building materials and sustainability

Some materials cause depletion during harvest or extraction, others give off pollutants during manufacture, emit pollutants into the occupied space, or contribute to solid waste problems at the end of their useful life (Wilson et al., 1998: 170). Furthermore, proper materials selection offers a type of quality control that can save money in remediation and lessen legal liability (Kibert, 2007b: 290–296). This reveals the importance of choosing sustainable materials. Materials should be selected:

- considering the following principles (Kibert, 2007a: 595–599):
  - closing materials loops by designing buildings for deconstruction and developing disassemblable building products with recyclable materials
  - the "cradle-to-cradle" approach: the extended version of the "cradle-to-grave" technique which includes the reuse, recovery and/or recycling – 3R – potential (Silvestre et al., 2014)
  - the environmental impact of materials extraction due to an increasing dependence on ever more dilute and distant stocks of ores and other resources and an increase in waste from the extraction of materials and the disposal of an ever growing built environment

- the need for project teams to deepen their interaction and the cross-sectorial implications of decision making;

- achieving closed-loop building materials via the following strategies (Kibert, 2007b: 245): buildings must be deconstructable and products must be disassemblable; materials must be recyclable. In addition, products/materials must be harmless in production and in use and then once recycled materials dissipated must be harmless reducing the life-cycle impacts associated with materials used in construction with the help of: locally sourced, sustainably harvested wood and easily recyclable materials or materials responsible for fewer pollutants (Wilson et al., 1998: 171);

- ensuring low environmental impact, issues of ongoing maintenance and durability throughout the building design and materials selection process (Wilson et al., 1998: 176);

- ensuring that the products and materials are harmless in use and in recycling (Kibert, 2007b: 274);

- using the integration of a BIM model with a decision-making tool for sustainable material selection so that options for improving the building materials' environmental sustainability performance are evaluated (Bank et al., 2011);

- keeping in mind the following three priorities in selecting building materials and products (Kibert, 2007b: 246):

  - reducing the quantity of materials needed for construction
  - reusing materials and products from existing buildings
  - using products and materials containing recycled content or made from renewable resources;

- evaluating the materials not only considering the LCA but also giving priority to the production and fate of materials and products, as the LCA fails to address closed-loop material behaviour (Kibert, 2007b: 274);

- considering the following recommendations (Kibert, 2007b: 290–296):

  - considering contaminants emitted (e.g. VOCs, e.g. water-based sealants manufactured using nontoxic components, particleboards manufactured with lower formaldehyde emissions, new products with zero or low emissions should be preferred for floor and wall coverings, water-based finishes, as well as insulation and ceiling tiles that do not contribute VOC and particulate contaminants should be chosen)
  - promoting the use of low-emissions building materials (including adhesives and sealants) (e.g. natural resins should be chosen instead of synthetic ones)
  - communicating IEQ requirements to subcontractors and suppliers (Kibert, 2007b: 290–296)
  - preparing the specifications based on the MasterFormat form of specifications developed by the Construction Specifications Institute.

It is important not to waste materials during construction. Material usage of a building can be improved via (Montoya, 2011: 94–98) reusing existing building stock or selecting building materials considering environmental, social and economic factors. In all cases companies should look towards recycling construction waste and reducing the overall amount of waste materials generated so that less waste is generated.

5.3.2 Building envelope

As we described at the beginning of the chapter, building energy consumption accounts for approximately 40% of the global energy demands (Wang et al., 2005; Dombaycı et al., 2006; Liu et al., 2008 and Wong and Mui, 2009 as quoted from Zheng et al., 2010: 710). Energy consumption can be reduced through (Zheng et al., 2010: 710): energy loss reduction; more energy-efficient building designs; improved thermal performance of building envelopes and successful design of building envelope. The building envelope, which separates the interior, conditioned environment from the exterior, unconditioned environment of a building, is the key determinant of thermal and energy performance in many types of buildings as it is primarily designed to restrict the heat transfer between inside and outside to regulate the thermal characteristics of the interior environment and reduce the heating, cooling and electric lighting demand of buildings (International Energy Agency, IEA, 2013). A building envelope with poorly insulated walls, roofs and foundations is typically characterised by up to 40% of total heat loss; drafts could result in up to 25% of total heat loss and low quality doors and windows could result in up to 30% of total heat loss (Hydro Q., 2008, as quoted from Zheng et al., 2010: 710). In the US building sector, 37% of primary energy consumption is due to space heating and cooling mainly caused by the building envelope characteristics, such as window/wall conductivity and infiltration rate (US Department of Energy, DOE, 2014). The design of the building envelope can significantly help achieve the heating and cooling objectives and improve energy efficiencies (Sozer, 2010). The designs of building envelopes have a significant impact on heat gain and cooling requirements (Wong et al., 2010). The improvement of the building envelope system causes a reduction of 55–65% (Schweiker and Shukuya, 2010: 2983). The building envelope determines the energy exchange between outdoor environment and indoor spaces and hence governs the overall energy performance of the building limiting thermal losses during winter and thermal gains during summer (Sozer, 2010: 2592). Building envelope design should be carried out taking into account several points.

First, thermal insulation, which is used to decrease heat transfer to/from surfaces, is one of the most effective energy conservation measures for cooling and heating in buildings, as lack of insulation causes larger life-cycle energy costs with lower initial capital (Bolattürk, 2008: 1055–1056). Appropriate thermal insulation, glazing type and shading elements can reduce the heat conducted through the building envelope (Sozer, 2010: 2592). A continuous insulation layer placed on the outdoor façade performs better than dispersed insulation within the building material with respect to reducing the cooling loads (Bolattürk, 2008: 1056). The optimum insulation thickness can be calculated considering the costs of insulation material and of energy, as well as cooling and heating loads, the efficiency of the heating system, the coefficient of performance of the cooling equipment, the lifetime of the building, as well as current inflation and discount rates (Bolattürk, 2008: 1056).

Second, the service life of the building envelope should be considered in economical assessments (Rudbeck, 2002: 83).

Third, the successful design of building envelopes system requires special attention in the conceptual stage (Zheng et al., 2010: 710). As summarised briefly by Zheng et al. (2010: 710–712), the following topics need to be taken into account while designing the building envelope:

- several building envelope alternatives should be evaluated to obtain the best solution, to overcome market barriers and to ensure incorporation of the cost effective energy efficiency opportunities into new buildings (Yu et al., 2009)
- the result of a systematic approach considering all relevant elements should be checked with respect to energy-related, environmental, capitalised and physical qualities (Bolin, 2008), environmental damage and risk to human health, with an emphasis on sustainable building envelope
- reliability of the building envelopes (Gowri, 1990).

Finally, Shen et al.'s (2011) study revealed that, depending on location, season and orientation, exterior and interior surface temperatures can be reduced by up to 20°C and 4.7°C, respectively, using different reflective coatings. Energy consumption can be reduced via (Shen et al., 2011: 1–8):

- light or white coloured surfaces (white or light coloured surfaces have lower surface temperature compared to dark coloured ones: Givoni and Hoffman, 1968; Bansal et al., 1992; Taha et al., 1992; Cheng et al., 2005; Synnefa et al., 2006; Uemoto et al., 2010)
- materials with small thermal resistance
- high reflective coatings where appropriate.

Technology and material choices of building envelopes have different environmental impacts (Pulselli et al., 2009: 920). For all orientations there is an important improvement in the energy performance of the building when designing according to the self-shading envelope (Capeluto, 2003: 327). Building envelopes should be designed for enhancing energetic performances relative to local climate conditions (Pulselli et al., 2009: 927). During the preliminary stages of a building's design, it is important that the architects deal with the solar potential of the building and the surrounding areas using sun diagrams, manual protractors or a computer and taking into account climatic considerations to assure, for example, the exposure of the elevations and sidewalks to the winter sun and to create the appropriate shading during the critical hours of the summer days (Capeluto, 2003: 327–328).

5.3.3 Building services engineering

The indoor environment of a building must provide proper thermal conditions to facilitate use of the space (Sozer, 2010). Buildings are expected to maintain a constant comfortable and healthy indoor climate with respect to variable external climate conditions (Pulselli et al., 2009: 920). Indoor environmental quality includes not only the subject of indoor air quality but also lighting quality, noise, temperature, humidity, odours and vibration (Kibert, 2007b: 277). A person's health, comfort and well-being can be affected by the quality of the indoor air (Montoya, 2011: 99). Indoor environment quality of a building can be improved via (Montoya, 2011: 99–103): mechanical ventilation systems, passive ventilation systems (i.e. stack ventilation) and hybrid ventilation systems (combination of mechanical and passive ventilation).

Indoor environmental quality is affected by (Kibert, 2007b: 281): operation and maintenance of the building; occupants of the building and their activities; building contents; outdoor environment and building fabric. Similarly, indoor air quality is affected by: the choice of materials and finishes in a building, activities like smoking and open combustion of gas- or wood-fired appliances, high moisture levels that permit mould growth and the amount (and quality) of fresh air introduced through ventilation (Wilson et al., 1998: 173–174).

About 30–40% of total energy consumption in Western countries is assigned to building and about half of this consumption referring to the energy consumption for indoor air conditioning (Bastianoni et al., 2007 – as quoted by Pulselli et al., 2009: 920). It is estimated that green buildings use an average of 36% less energy than conventional buildings and that they have corresponding reductions in CO_2 emissions (Montoya, 2011: 29). If heating and cooling loads are kept low through careful envelope design, glazing selection and lighting design, heating and cooling equipment can be significantly downsized or, in some cases, even eliminated (Wilson et al., 1998: 163). Energy efficiency of a building can be improved via (Montoya, 2011: 61–93): heating and cooling systems, passive solar heating and cooling, maximising energy performance, using renewable energy sources and on-site power generation using renewable energy sources.

The energy efficiency in the big scale buildings cannot only be improved by the application of advanced mechanical systems and advanced technologies, but also by the design decisions that affect operation and management (Sozer, 2010: 2582). For this reason, it is important to carefully design the HVAC (heating, ventilation and air conditioning) system. The HVAC system must (Kibert, 2007b: 287–289):

- provide the proper balance of temperature (65–78°F or 18–26°C) and humidity (30–60%) to maintain a comfortable indoor environment
- be capable of controlling the supply air in spite of changing conditions in the return of outside sources
- control moisture
- be designed to ensure that:
  - air intakes are located away from exhaust sources
  - air intakes are high enough above the ground to avoid bringing in ground source contaminants
  - proper zoning is carried out to prevent unnecessary circulation of internal contaminants.

In addition to the considerations on material selection, building envelope and HVAC, it is important to take into account the following points as well:

- *Building life span*: As buildings typically have a long life span, lasting for 50 years or more, it is important to be able to analyse how buildings will respond to climate change in the future and to assess the likely changes in energy use (Wong et al., 2010).

- *Lighting*: Lighting should be based on the following principles (Wilson et al., 1998: 158):

    ○ Lighting from multiple sources, optimum light levels for comfort, minimum glare, appropriate levels of contrast between light and dark surfaces and aesthetics need to be considered for high-quality lighting.
    ○ Natural lighting can be provided with the help of light scoops, skylights and high-performance glazings.
    ○ Sun-tracking lighting system can be installed on the roof.
    ○ Daylight can be brought deep into the building.

- *Acoustics*: Building type, shape, number of storeys, relative location to noise sources, as well as materials for external and internal envelopes affect the acoustics performance of the building (Yu and Kang, 2009: 2166). The environmental impacts of acoustic materials should be taken into account at various stages of the design process, from initial planning to interior design and correspondingly, the results are relevant to different users, including architects, acousticians and planners (Yu and Kang, 2009: 2175). The concerns regarding the effects of acoustical ceiling tiles on indoor air quality include: the occurrence of microbial growth on either mineral fibre or fibreglass tile exposed to moisture; porous tile absorption and re-emission of VOCs (Kibert, 2007b: 296).

- *Planting*: Planting climatically appropriate vegetation can eliminate the need for irrigation and chemical herbicides and insecticides, reducing maintenance costs (Wilson et al., 1998: 150).

- *Water use efficiency*: Water use efficiency of a project can be improved via (Montoya, 2011: 53–56): water-efficient landscaping (i.e. choosing plant species tolerant to the specific climate or microclimate), water-efficient buildings (i.e. using low-flow plumbing fixtures) and recycling wastewater (i.e. reusing rainwater runoff and greywater for irrigation).

5.4 Conclusion

The world's habitat is deteriorating at an accelerated rate especially due to the overconsumption of natural resources. The construction industry plays an important role in the exploitation of the world's resources. The built environment has a large negative effect on ecosystem services (Reed, 2007; Graham, 2003) and it is increasingly held accountable for global environmental and social problems (Zari, 2012: 57–62). According to the United Nations Environment Program (UNEP Sustainable Buildings and Construction Initiative, 2007), 40% of all energy and material resources are used to build and operate buildings globally. The global pollution attributable to buildings includes (Brown and Bardi, 2001): deterioration of air quality (23%); emission of greenhouse gases (50%); pollution of drinking water (40%); depletion of ozone (50%); and causing landfill waste (50%). Living in harmony with nature is the key for the next generations to survive and for sustaining nature. For this reason, this chapter has focused on sustainable buildings outlining the definition of and drivers for sustainable buildings, building assessment tools and key technical aspects of sustainable buildings. Whereas the focus of green buildings is all about reducing the negative impacts of the buildings to the environment, the focus of sustainable buildings is on causing zero negative impact to the environment. The new trend in the near future will be the shift from sustainable buildings towards regenerative buildings, which aim for net positive impact on the environment (i.e. producing not only its own energy but also supplying energy to the vicinity). This new concept, which is in its infancy, will require new technologies, new materials, new concepts, new management processes and new building assessment tools to be invented.

The next chapter introduces low and zero carbon technologies used in sustainable buildings as these technologies support minimisation of carbon emissions from buildings.

References

Building assessment tools

Baldwin, R., Leach, S.J., Doggart, J. and Attenborough, M. (1990). BREEAM 1/90: An Environmental Assessment for New Office Designs. Building Research Establishments, Garston.

Baldwin, R., Yales, A. and Rao, S. (1998). BREEAM 98 for Offices: An Environmental Assessment Method for Office Buildings, Centre for Sustainable Construction, Building Research Establishment, Watford, UK.

BRE Global. (2009). BREEAM BRE Environmental and Sustainability Standard BES 5051 Issue 3.0. BREEAM education 2008 Assessor Manual. (Available at http://www.leakdetectionuk.com/resource_centre/liquid_leak_detection/downloads/BREEAM_EDUCATION_2008.pdf, accessed 1 July 2015).

Burnett J. (2007). City building – Eco-labels and shades of green. Landscape and urban planning, 83, 29-38.

CASBEE website http://www.ibec.or.jp/CASBEE/english/index.htm

CASBEE (2008a). CASBEE for New Construction – Technical Manual 2008 Edition. Tool-1.

CASBEE (2008b). CASBEE for an Urban Area + Buildings Comprehensive Assessment System for Building Environmental Efficiency. Technical Manual 2007 Edition. Tool-21+.

Centre for Environmental Technology (CET) (1996), HK-BEAM 1/96 (New Offices). An Environmental Assessment for New Air-conditioned Office Premises, Hong Kong.

Chau, C.K., Lee, W.L., Yik F.W.H. and Burnett, J. (2000). Towards a Successful Voluntary Building Environmental Assessment Scheme. Construction Management and Economics, 18(8), 959–968.

Cole, R. (1998). Emerging Trends in Building Environmental Assessment Methods. Building Research and Information, 26, 3–16.

Cole, R.J. (2003). Building environmental assessment methods: a measure of success. Int. Electr. J. Construct. Future Sustain. Construct., 1–8.

Crawley, D. and Aho, I. (1999). Building Environmental Assessment Methods: Applications and Development Trends. Building Research and Information, 27(4/5), 300–308.

GBCA (2005). Green Building Council of Australia Environmental Rating System for Buildings. www.gbcaus.org. Certification Trade Mark Rules. Trade Mark No. 960850.

GBCA (2008). The Green Building Council Australia. Building Sustainable Future. 12 November 2008.

GBCA (2009a). Green Building Council Australia. Green Star. Building a Sustainable Future. Updated: 1 May 2009. Green Star Certification Process 1–7, Sydney.

GBCA (2009b). Green Building Council Australia. Green Star. Green Star Eligibility Criteria. Updated 1 April 2009. Building Sustainable Future, Sydney.

Haapio, A. and Viitaniemi, P. (2008). A Critical Review of Building Environmental Assessment Tools. Environmental Impact Assessment Review, 28, 469–482.

HK-BEAM Society (2004). HK-BEAM 4/04 New Buildings. (Available at: http://www.hkbeam.org.hk/, accessed 30 July 2015).

Hueting, R. and Reijnders, L. (2004). Broad Sustainability Contra Sustainability: The proper construction of sustainability indicators. Ecological Economics, 50(3–4), 249–260.

Institute of Building Environment and Energy Conservation (2003). (Available at http://www.ibec.or.jp/aboutus.html, accessed 30 July 2015).

Japan Sustainable Building Database (Available at http://www.ibec.or.jp/jsbd/, accessed 30 July 2015).

Saparauskas, J. (2007). The main aspects of sustainability evaluation in construction. In: 9th International Conference on Modern Building Materials, Structures and Techniques, Vilnius, Lithuania.

Saunders, T. (2008). A Discussion Document Comparing International Environmental Assessment Methods for Buildings. (Available at http://www.breeam.org/filelibrary/International%20Comparison%20Document/Comparsion_of_International_Environmental_Assessment_Methods01.pdf, accessed 30 July 2015).

U.S. Green Building Council (1999). LEEDTM—Leadership in Energy and Environmental Design.

U.S. Green Building Council (2003). Green Building Rating System For New Construction & Major Renovations (LEED-NC) Version 2.1.

Building envelope

Bansal, N.K., Garg, S.N. and Kothari, S. (1992). Effect of Exterior Surface Colour on the Thermal Performance of Buildings. Building and Environment, 27, 31–37.

Bolattürk, A. (2008). Optimum Insulation Thicknesses for Building Walls with Respect to Cooling and Heating Degree-Hours in the Warmest Zone of Turkey. Building and Environment, 43, 1055–1064.

Bolin, R. (2008). Sustainability of the Building Envelope. (Available at http://www.wbdg.org/resources/env_sustainability.php?r=envelope, accessed 7 July 2015).

Capeluto, I.G. (2003). Energy Performance of the Self-Shading Building Envelope. Energy and Buildings, 35, 327–336.

Cheng, V., Ng, E. and Givoni, B. (2005). Effect of Envelope Colour and Thermal Mass on Indoor Temperatures in Hot Humid Climate. Solar Energy, 78, 528–534.

DOE (2014). Windows and Building Envelope Research and Development: Roadmap for Emerging Technologies. US Department of Agency.

Dombaycı, Ö., Gölcü, M. and Pancar, Y. (2006). Optimization of Insulation Thickness for External Walls using Different Energy-Sources. Applied Energy, 83, 921–928.

Givoni, B. and Hoffman, M.E. (1968). Effect of Building Materials on Internal Temperatures, Research Report, Building Research Station, Technion Haifa.

Gowri, K. (1990). Knowledge-Based System Approach to Building Envelope Design. Ph.D. Thesis, Centre for Building Studies, Concordia University, Montreal, Canada.

International Energy Agency (2013). https://www.iea.org/publications/freepub lications/publication/2013_AnnualReport.pdf

Liu, Y., Joseph, C.L. and Tsang, C.L. (2008). Energy Performance of Building Envelopes in Different Climate Zones in China. Applied Energy, 85, 800–817.

Pulselli, R.M., Simoncini, E. and Marchettini, N. (2009). Energy and Energy Based Cost–Benefit Evaluation of Building Envelopes Relative to Geographical Location and Climate. Building and Environment, 44, 920–928.

Rudbeck, C. (2002). Service Life of Building Envelope Components: Making it Operational in Economical Assessment. Construction and Building Materials, 16, 83–89.

Schweiker, M. and Shukuya, M. (2010). Comparative Effects of Building Envelope Improvements and Occupant Behavioural Changes on the Exergy Consumption for Heating and Cooling. Energy Policy, 38, 2976–2986.

Shen, H., Tan, H. and Tzempelikos, A. (2011). The Effect of Reflective Coatings on Building Surface Temperatures, Indoor Environment and Energy Consumption – An Experimental Study. Energy and Buildings, 43(2–3), 573–580.

Sozer, H. (2010). Improving Energy Efficiency through the Design of the Building Envelope. Building and Environment, 45, 2581–2593.

Synnefa, A., Santamouris, M. and Livada, I. (2006). A Study of the Thermal Performance of Reflective Coatings for the Urban Environment. Solar Energy, 80, 968–981.

Uemoto, K.L. Neide, S.M.N. and Vanderley, J.M. (2010). Estimating Thermal Performance of Cool Colored Paints. Energy and Buildings, 42(1), January 2010, 17–22.

Wong, S.L., Wan, K.K.W., Li, D.H.W. and Lam, J.C. (2010). Impact of Climate Change on Residential Building Envelope Cooling Loads in Subtropical Climates. Energy and Buildings, 42(11), 2098–2103.

Yu, J.H., Yang, C.Z., Tian, L.W. and Dan, L. (2009). Evaluation on Energy and Thermal Performance for Residential Envelopes in Hot Summer and Cold Winter Zone of China. Applied Energy, 86, 1970–1985.

Zheng, G., Jing, Y., Huang, H. and Gao, Y. (2010). Application of Improved Grey Relational Projection Method to Evaluate Sustainable Building Envelope Performance. Applied Energy, 87, 710–720.

Economic assessment/life cycle assessment

Bank, L.C., Thompson, B.P. and McCarthy, M. (2011). Special report: Decision-Making Tools for Evaluating the Impact of Materials Selection on the Carbon Footprint of Buildings. Carbon Management, 2(4), 431–441.

Bastianoni, S., Galli, A., Pulselli, R.M. and Niccolucci, V. (2007). Environmental and Economic Evaluation of Natural Capital Appropriation through Building Construction: Practical Case Study in the Italian Context. Ambio, 36(7), 559–565.

Bisset, R. (1980). Methods for Environmental Impact Analysis: Recent Trends and Future Prospects. Journal of Environmental Management, 11, 27–43.

Ding, G.K.C. (2005). Developing a Multicriteria Approach for the Measurement of Sustainable Performance. Building Research and Information, 33(1), 3–16.

Finch, E. (1992). Environmental Assessment of Construction Projects. Construction Management and Economics, 10, 5–18.

IEA International Energy Agency (2004). Life Cycle Assessment Methods for Buildings, Annex 31-"Energy Related Environmental Impacts of Buildings", Canada.

Levin, H. (1997). (Hal Levin & Associates, Santa Crux, CA) Systematic Evaluation and Assessment of Building Environmental Performance (SEABEP). In: Second International Conference Buildings and Environment, June 1997, Paris, pp. 3–10.

RICS (2001). Comprehensive Project Appraisal: Towards Sustainability. RICS Policy Unit, RICS, London.

Silvestre, J.D., Brito, J. and Pinheiro, M.D. (2014). Environmental impacts and benefits of the end-of-life of building materials – calculation rules, results and contribution to a "cradle to cradle" life cycle. Journal of Cleaner Production 66, 37–45.

Tatari, O. and Kucukvar, M. (2010). Cost Premium Prediction of Certified Green Buildings: A Neural Network Approach. Building and Environment, 46(5), 1081–1086. doi:10.1016/j.buildenv.2010.11.009

Tisdell, C. (1993). Project Appraisal, the Environment and Sustainability for Small Islands. World Development, 21(2), 213–219.

Vatalis, K.I., Manoliadis, O.G. and Charalampides, G. (2011). Assessment of the Economic Benefits from Sustainable Construction in Greece. International Journal of Sustainable Development and World Ecology, 18(5), 377–383.

Wang, W.M., Radu, Z. and Hugues, R. (2005). Applying Multi-Objective Genetic Algorithms in Green Building. Build Environment, 40, 1512–1525.

Wehrmeyer, W. and Tyteca, D. (1998). Measuring Environmental Performance for Industry: From Legitimacy to Sustainability and Biodiversity? International Journal for Sustained Development of World Ecology, 5, 111–124.

Wilson, A., Uncapher, J.L., McManigal, L., Lovins, L.H., Cureton, M. and Browning, W.D. (1998). Green Development Integrating Ecology and Real Estate. John Wiley & Sons, Inc., New York.

Wong, L.T. and Mui, K.W. (2009). Efficiency Assessment of Indoor Environmental Policy for Air-Conditioned Offices in Hong Kong. Applied Energy, 86, 1933–1938.

Energy efficiency

Brown, M.T. and Bardi, E. (2001). Handbook of Energy Evaluation. A Compendium of Data for Energy Computation Issued in a Series of Folios. Folio No. 3: Energy of Ecosystems. Center for Environmental Policy, Environmental Engineering Sciences, University of Florida, Gainesville. (Available at http://www.epa.gov/aed/html/collaboration/emergycourse/presentations/Folio3.pdf , accessed 30 July 2015).

EIA (Energy Information Administration). (2008). Annual Energy Outlook. US Department of Energy. (Available at http://www.eia.doe.gov/oiaf/aeo/index.html, accessed 7 July 2015).

Hobbs, B.F. and Meier, P. (2000). Energy Decision and the Environment: A Guide to the Use of Multicriteria Methods. Kluwer, Boston, MA.

Taha, H., Sailor, D. and Akbari, H. (1992). Highalbedo Materials for Reducing Cooling Energy Use. Lawrence Berkeley Laboratory Report 31721, UC-530, Berkley, CA.

Passive design

Rodriguez-Ubinas, E., Montero, C., Porteros, M., Vega, S., Navarro, I., Castillo-Cagigal, M., Matallanas, E. and Gutiérrez, A. (2014). Passive Design Strategies and Performance of Net Energy Plus Houses. Energy Buildings. http://dx.doi.org/10.1016/j.enbuild.2014.03.074.

Whang, S.W. and Kim, S. (2014). Determining Sustainable Design Management using Passive Design Elements for a Zero Emission House During the Schematic Design. Energy and Buildings, 77, 304–312.

Regenerative construction

Reed, B. (2007). Shifting from "Sustainability" to Regeneration. Building Research and Information, 35(6), 674–680.

Zari, M.P. (2012). Ecosystem Services Analysis for the Design of Regenerative Built Environments. Building Research and Information, 40(1), 54–64.

Sustainable construction project management

Fernández-Sánchez, G. and Rodríguez-López, F. (2010). A Methodology to Identify Sustainability Indicators in Construction Project Management – Application to Infrastructure Projects in Spain. Ecological Indicators, 10, 1193–1201.

Glavinich, T.E. (2008). Contractor's Guide to Green Building Construction Management, Project Delivery, Documentation and Risk Reduction, John Wiley & Sons, Hoboken, NJ.

Hydro Q. (2008). What Influences Your Design? (Available at: http://www. hydroquebec.com/residential/, accessed 30 July 2015).

Lam, P.T.I., Chan, E.H.W., Poon, C.S., Chau, C.K. and Chun, K.P. (2010). Factors Affecting the Implementation of Green Specifications in Construction. Journal of Environmental Management, 91(3), 654–661.

Yudelson, J. (2008). The Green Building Revolution. Island Press, Washington.

Sustainable construction

Alarcón Núñez, D.B. (2005). Modelo Integrado de Valor para Estructuras Sostenibles. Thesis. Universitat Politècnica de Catalunya, Escola Tècnica Superior D'Enginyers de Camins, Canals i Ports, Spain.

Bon, R. and Hutchinson, K. (2000). Sustainable Construction, Some Economic Challenges. Building Research & Information, 28(5/6), 310–314.

CIB (1999). Agenda 21 on Sustainable Construction. CIB, Rotterdam, The Netherlands.

DETR (Department of the Environment, Transport and the Regions) (2000). Building a Better Quality of Life: A Strategy for More Sustainable Construction. DETR, London, UK.

Doughty, M.R.C. and Hammond, G.P. (2004). Sustainability and the Built Environment at and beyond the City Scale. Building and Environment, 39 (10), 1223–1233.

Du Plessis, C. (2007). A Strategic Framework for Sustainable Construction in Developing Countries. Construction Management and Economics, 25(1), 67–76.

Fernández Sánchez, G. (2008). Análisis de los Sistemas de Indicadores de Sostenibilidad. Planificación urbana y proyectos de construcción, Escuela Técnica Superior de Ingenieros de Caminos, Canales y Puertos, Universidad Politécnica de Madrid, Spain.

Graham, P. (2003) Building Ecology – First Principles for a Sustainable Built Environment, Blackwell, Oxford.

Hill, R.C. and Bowen, P.A. (1997). Sustainable Construction: Principles and a Framework for Attainment. Construction Management and Economics, 15, 223–239.

Huang, R.Y. and Hsu, W.T. (2011). Framework Development for State Level Appraisal Indicators of Sustainable Construction. Civil Engineering and Environmental Systems, 28(2), 143–164.

Kibert, C.J. (1994). Principles of Sustainable Construction. In: Proceedings of the First International Conference on Sustainable Construction; 6–9 November 1994; Tampa, FL, pp. 1–9.

Kibert, C.J. and Grosskopf, K. (2005). Radical sustainable construction: envisioning next-generation green buildings. (Available at: http://www.cce.ufl. edu/wp-content/uploads/2012/08/WhitePaper-RSC06.pdf, accessed 30 July 2015).

Kibert, C.J. (2007a). The Next Generation of Sustainable Construction. Building Research and Information, 35(6), 595–601.

Kibert, C.J. (2007b). Sustainable Construction Green Building Design and Delivery, 2nd ed. John Wiley & Sons, Inc., Hoboken, NJ.

Montoya, M. (2011). Green Building Fundamentals. Practical Guide to Understanding and Applying Fundamental Sustainable Construction Practices and the LEED System, 2nd ed. Prentice Hall, Upper Saddle River, NJ.

Reed, B. (2007). Shifting from Sustainability to Regeneration. Building Research & Information, 35(6), 674–680.

Rodríguez-López, F. and Fernández-Sánchez, G. (2008). Sustainability and climate change: new objectives and requirements in Engineering Project Management. In: XII International Conference on Project Engineering AEIPRO-IPMA, Zaragoza, Spain.

Seo, S., Aramaki, T., Hwang, Y. and Hanaki, K. (2004). Fuzzy Decision-making Tool for Environmental Sustainable Buildings. Journal of Construction Engineering and Management (May/June), 415–423.

United Nations Environment Program (UNEP)–Sustainable Buildings and Construction Initiative (2007) Buildings and Climate Change: Status, Challenges and Opportunities, UNEP, Paris.

Yu, C.J. and Kang, J. (2009). Environmental Impact of Acoustic Materials in Residential Buildings. Building and Environment, 44, 2166.

6   Low- and Zero-Carbon Technologies in Buildings

Laurence Brady

6.1 Introduction

Low- and zero-carbon technologies used in buildings are renewable or highly efficient systems that can generate all or some of the energy needed to operate a building. This chapter will consider four of these technologies: combined heat and power, heat pumps, wind turbines and solar hot water.

Within the European Union buildings use 40% of the energy and produce 36% of Europe's carbon emissions. Most of the energy use in buildings occurs during the operational phase. In 2004, the European Energy Performance of Buildings Directive (EPBD) came into force. The purpose of this legislative instrument was to encourage the improvement in building energy performance in member states (*European Journal*, 2003).

Low- and zero-carbon technologies can play an important role, but only after designers and builders ensure that the building energy demand is minimised (Bell, 2013). This minimisation of energy demand should be achieved by designing and constructing a building that is well insulated, not leaky, exploits passive energy opportunities and contains engineering services that have been commissioned to obtain optimum performance. Additionally, the building operator will have the necessary knowledge and controls to be able to operate the building at, or somewhere near, that optimum performance level.

The successful implementation of low- and zero-carbon technologies requires that they be considered as part of an overall strategy. A carbon hierarchy has been developed that can provide a basic strategy. By adopting a carbon hierarchy, designers first of all make sure that building designs are as efficient as possible so as to minimise the buildings' carbon footprint. Only then are renewables considered (Zero Carbon Hub, 2014). Figure 6.1 suggests the roadmap to efficient building.

Figure 6.1 shows that fabric efficiency is the base measure. The building structure should exploit passive measures to reduce energy demand. For example, intelligently-designed thermally heavy construction can absorb heat and reduce the need for air conditioning. Passive environmental control measures are those methods that architects and builders used before technologies such as air conditioning were available. By careful design of structures and intelligent selection of materials, building designers set out to use the thermal inertia of the building fabric to store heat or limit the build-up of internal temperatures. Similarly, earlier building designs encouraged natural ventilation. They also designed buildings that would encourage natural ventilation.

Good design and workmanship should ensure that buildings are properly insulated and are sufficiently airtight. Achieving an efficient building fabric means that the engineering services need only be designed and sized to meet the loads that cannot be met passively. Designers should select efficient equipment but also

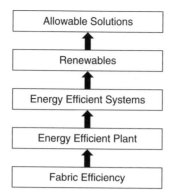

Figure 6.1 Roadmap to efficient building

ensure that the engineering services can be operated efficiently. Optimum performance for mechanical and electrical engineering systems requires that operators can understand and maintain these services. Engineering plant that cannot be maintained will not operate at optimum efficiency. When fabric and plant efficiencies have been achieved, the situation is then set for successful introduction of renewable technologies (CIBSE, 2012). Combined heat and power, heat pumps, wind turbines and solar hot water are presented in the following sections.

6.2 Combined heat and power

Electricity is a versatile and convenient fuel. Electricity can provide the power to heat or cool buildings. It provides the energy to illuminate buildings, power lifts, energise computers and much more. This energy source is essential to our modern way of life. Most of us live and work in buildings in which electricity is supplied from power stations which supply large distribution grids. Traditional grid-supplied electricity is mostly generated in central power stations that operate at thermal efficiencies of perhaps 40%. Combined cycle power stations, where electricity generators are driven by gas turbines whose exhaust gases are used to drive a steam turbine, achieve efficiencies of around 50%. The location of power stations often means that there are also losses in transmitting energy from power station to source.

Electricity comes at a primary energy cost. 50% of the primary energy delivered to a power station can be exhausted as waste heat. Reference to any thermodynamics text book will reveal that the maximum theoretical efficiency of a heat engine cycle is given by (1):

$$\eta = 1 - \frac{T_1}{T_2} \tag{1}$$

where η is efficiency, T_2 is the absolute temperature of high temperature energy input reservoir and T_1 is the absolute temperature of the low temperature energy reservoir. The example that follows illustrates how ideal efficiencies can be determined from system temperatures.

Example 1.

Question

Determine the efficiency for a heat engine system that is supplied with steam at 2000°C. The steam is exhausted and condensed at a temperature of 30°C.

Answer

$$\eta = 1 - \frac{(273+30)}{(273+2000)} = 86.7\%$$

Thermal inefficiency in conventional power generation is an inescapable fact. The efficiency formula given in (1) is based on natural laws and systems. The practical way to improve the efficiency of the process is to make use of the heat that is normally rejected into the atmosphere.

A system that generates electricity and makes use of the heat rejected from the process is a "Combined Heat and Power (CHP) System". It is important to remember that the efficiency of the process is only really increased if the heat and power produced provide useful energy. A typical application where the heat can be used is in a district heating arrangement where numerous dwellings can be supplied with energy for heating or hot water. On a smaller scale, a CHP plant may be designed to provide heat and power to a specific building.

Power plants use different prime movers to drive generators. In large power stations, steam or gas turbines will drive the electric generators. Gas turbines may also be used on smaller projects. Reciprocating engines are commonly used for small scale CHP units for buildings, often fuelled by natural gas. Domestic sized CHP units have been developed that use sterling engines. An important characteristic of prime movers is the ratio of heat to electrical power produced. The example that follows illustrates how a small scale CHP system can reduce carbon emissions.

Example 2.

Question: A gas fuelled CHP unit with an electrical output of 80 kW and a heat output of 149 kW operates for 16 hours/day. The engine has a mechanical efficiency of 35%. The plant runs 7 days/week for 48 weeks of the year. Determine the carbon saving/year if the CHP unit replaces a gas-fired boiler (78% efficiency).

Answer:

Natural gas	0.1836 kg CO_2/kWh	
Grid electricity	0.5246 kg CO_2/kWh	
Boiler		
Gas input to boiler	= 149/0.78	= 191 kW
	= 191 × 16 × 7 × 48	= 1,026,816 kWh/year
Carbon emissions	= 1,026,816 × 0.1836	= 188,523 kg/year
Grid electricity	= 80 × 16 × 7 × 48	= 430,080 kWh/year
Carbon emissions	= 430,080 × 0.5246	= 225,620 kg/year
Total Carbon emissions		= 414,143 kg/year

CHP

Gas input to engine	$= 80/0.35$	$= 229$ kW
Heat output from engine	$= 229 \times (1 - 0.35)$	$= 149$ kW
Carbon emissions	$= 229 \times 16 \times 7 \times 48 \times 0.1836$	$= 226{,}031$ kg/year
Carbon savings	$= 414{,}143 - 226{,}031$	$= 188{,}112$ kg/year

This simple example assumes that the heat to power ratio of the prime mover provides the exact amount of heat and power required. An ideal CHP installation in a building is one in which the heat and power produced by the engine/generator exactly meets the electricity and heat demand of the building every day throughout the year. Given that electrical and heating loads vary daily and seasonally, selecting a CHP prime mover with a heat-to-power ratio that can meet these varying loads is clearly not a practical possibility. Sizing a CHP system for a building is therefore critical to its success.

The sizing strategy will differ for different applications. Some industrial applications for CHP have large, year round heat loads. For many buildings much of the heat load is seasonal and a CHP unit sized to meet the winter heat load may produce too much heat in summer. This could mean that heat energy is produced for which there is no use and would need to be "dumped" to atmosphere.

Where a CHP generates more electricity than is required it may be possible to return electric power back to the grid. This can be more attractive financially if a "feed in tariff" arrangement exists. In other words, excess electrical power generated by the CHP system could be "sold" back into the grid.

In some cases CHP units are sized so that they contribute a fraction of the heating and electrical loads for a building. Although this means that buildings may also require heating plant, and may require to draw electricity from the grid, the energy from the CHP will always be required and, therefore, the CHP should operate at optimum overall thermal efficiency.

Figure 6.2 illustrates a CHP unit arranged in parallel with boiler plant. If the CHP was suitably rated and acted as "lead boiler", then its heat energy should

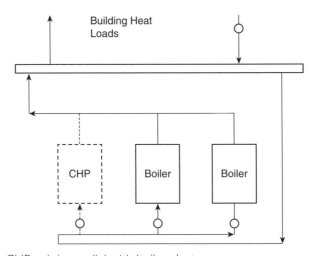

Figure 6.2 CHP unit in parallel with boiler plant

always be put to use. If the heat load becomes greater than the capacity of the CHP unit, then the other boilers can fire.

Incorporating CHP plant to work alongside boiler plant and grid-supplied electricity will need careful design and control. Building plant will become more complicated with associated maintenance needs.

Although CHP is not a renewable technology CHP systems are capable of providing significant amounts of heat and electricity, whilst not requiring impractical amounts of plant space. For example, some large general hospitals in the UK have retro-fitted CHP units to replace boiler plant and have been able to locate the CHP units in the spaces previously occupied by the boiler plant.

6.3 Heat pumps

Like CHP, heat pumps are not actually renewable energy systems. Heat pumps are normally powered by electricity, which, if grid supplied, is generated at poor efficiencies. However, the ratio of energy output to input for heat pumps can be high and this means that they can reduce energy demand in buildings.

Most people are familiar with the refrigerator in their kitchen that keeps their milk and cheese cool. The process of keeping the inside compartment of the refrigerator cool is actually one of extracting the heat from this compartment. The extracted heat is rejected to the space surrounding the refrigerator. If the role of the refrigerator apparatus is to heat the space around the refrigerator rather than cool the internal compartment, this equipment is called a "heat pump". Most heat pumps are vapour compression units. The major components are shown in Figure 6.3.

Evaporator: A liquid (refrigerant) is introduced into the evaporator. Because the evaporator is at a low pressure, the liquid evaporates. This process is similar

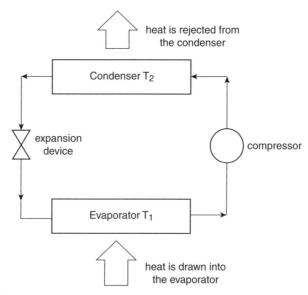

Figure 6.3 Major components of heat pump

to what happens to the liquid butane used in camping stoves. When the valve is opened, the pressurised liquid evaporates as it is exposed to the lower atmospheric pressure. Notably, camping stoves burn gas and not liquid. The evaporation process draws heat from the surroundings and cools accordingly.

Compressor: The refrigerant gas has its pressure raised by the compressor, which also circulates the refrigerant around the system.

Condenser: Having had its pressure raised, the gas can now be condensed back into a liquid in the condenser.

Expansion device: The high pressure liquid enters the expansion device, which reduces its pressure so that it can re-evaporate in the evaporator.

The vapour compression refrigeration cycle operates as a reverse cycle heat engine. This is because the role of the refrigeration cycle is not to use heat to generate power, but to use power to transfer heat from a cold source to a hot source. If the purpose of an apparatus is to provide cooling, it is called a refrigerator. If the purpose of an apparatus is to provide heating, then it is called a heat pump.

The performance of a heat pump is described by the term "Coefficient of Performance" or COP. The COP is the ratio of the heat output to the energy input. In an ideal theoretical situation, the only energy input is that required to drive the compressor. The ideal theoretical value for Coefficient of Performance for a heat pump may be found from (2):

$$COP = \frac{T_2}{(T_2 - T_1)} \tag{2}$$

where T_1 is the absolute evaporator temperature and T_2 is the absolute condenser temperature.

The example that follows demonstrates how a smaller temperature difference between the evaporator and the condenser can increase the COP of a heat pump.

Example 3.

Question

Determine the Coefficient of Performance for a heat pump operating with a condenser temperature of 35°C and an evaporator temperature of 0°C.

Answer

$$COP = \frac{(273 + 35)}{((273 + 35) - 273)} = 8.8$$

This theoretical value for COP is high and would not be achieved in practical systems because the processes would not be ideal and there would be other inefficiencies. However, in the theoretical example, it may be seen that 1 kW of energy input would provide 8.8 kW of heating. Actual systems may operate with COPs of perhaps between 2 and 4. Nevertheless, this ideal formula is useful because it demonstrates that the temperature difference between the evaporator and the condenser affects the performance of the heat pump. Greater COPs are achieved if this temperature difference is reduced.

In practical terms this means that higher COPs are achieved if the condenser is operated at a higher temperature. In other words, the temperature of the source from which the heat is extracted will affect the COP. This can be seen if an air source heat pump operating in winter is compared to a ground source heat pump.

Example 4 provides a practical demonstration of why it can be more economical to run a ground source heat pump compared to an air source heat pump.

Example 4.

Question

A building is to be heated by a heat pump during winter. Heating load is 25 kW. Outside air temperature is –5°C and the ground temperature is 10°C. If the heat pump with 60% efficiency supplies an underfloor heating system with heat at 40°C, determine the energy input if the heat pump is:

a.   An air source heat pump;
b.   A ground source heat pump.

Answer a.

Air source

$$COP = \frac{T_2}{(T_2 - T_1)} = \frac{(40+273)}{((40+273)-(273-5))} = 6.95 \times 0.6 = 4.17$$

Energy input = 25/4.17 = 6 kW

Answer b.

Ground source

$$COP = \frac{T_2}{(T_2 - T_1)} = \frac{(40+273)}{((40+273)-(273-10))} = 10.43 \times 0.6 = 6.25$$

Energy input = 25/6.25 = 4 kW

From Example 4, it can be seen that a higher temperature heat source improves the Coefficient of Performance. Ground source heat pumps, as shown in Figure 6.4, can provide higher COPs because, at 10–15 m depth, the temperature of the ground can remain at about 10–12°C throughout the year. However, the civil engineering work required to embed the ground source heat exchanger can be extensive and it is important to examine the condition of local ground to assess its heat transfer properties.

6.4 Wind turbines

The energy available in wind is due to its motion. Kinetic energy contained in wind flow is a function of the mass of air flow m and its velocity V as expressed by (3).

Kinetic Energy $= 0.5 \times m \times V^2$ $\hspace{2cm}$ (3)

Figure 6.4 Boiler and heat pump assembly for under-floor heating

It is simpler to visualise air flow in volumetric terms, as shown in Figure 6.5, rather than mass terms. Consider a volume of air approaching a horizontal axis wind turbine in which the swept area covered by the blades is equal to πr^2, where r is the length of turbine blade. The air flow approaching the turbine can be considered in terms of a tubular volume. The cross sectional area of this "tube" of air will be equal to the swept area of the turbine blades and its length will be equal to the wind speed.

The following Examples (5, 6 and 7) illustrate how varying wind speed and the cross sectional of the swept area of turbine blades can have a significant effect on turbine power outputs.

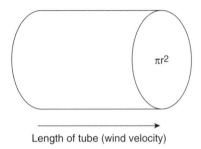

Length of tube (wind velocity)

Figure 6.5 Visualisation of air flow

Example 5.

Question

Determine the volume and mass of air approaching a horizontal axis wind turbine, if the wind speed is 3 m/s and the blade radius is 10 m.

Answer

Volume flow of air = $(\pi r^2 \times V) = \pi \ 10^2 \times 3 = 942.5$ m³/s

If air is taken to have a density (ρ) of 1.2 kg/m³, then the volume flow of air can be converted if it is multiplied by 1.2.

Mass flow of air = $942.3 \times 1.2 = 1131$ kg/s.

Example 6.

Question

Determine the energy contained in the wind approaching a 10 m radius horizontal axis wind turbine if the wind speed is 3 m/s.

Answer

Kinetic Energy = $0.5 \times 1131 \times 3^2 = 5090$ W

It is convenient to combine the expressions for kinetic energy and mass rate:

$0.5 \times (\pi r^2 \times V \times \rho) \times V^2$

Energy in approaching wind = $0.5 \ \pi \ \rho \ r^2 \ V^3$

It can be seen from this formula that the energy in wind is proportional to the cube of the wind speed.

Example 7.

Question

Determine the energy in wind approaching a horizontal axis wind turbine if the turbine radius is 16.5 m, wind speed is 12 m/s, and the density of air is 1.2 kg/m³.

Answer

Energy in approaching wind = $0.5 \times \pi \times 1.2 \times 16.5^2 \times 12^3 = 886,774$ W

The previous section examined how to assess the power contained in free-flowing wind. The energy extracted from that wind by a turbine is always less. A wind turbine develops power by extracting kinetic energy from the wind. This can be determined from the change in wind speed before and after the wind passes through the turbine according to (4):

Power output from turbine = $0.5 \times m \times (V_1^2 - V_2^2)$ (4)

A wind energy device cannot extract all of the kinetic energy in the wind since this would require the wind velocity to change to zero. In other words, the wind would have to stop moving. This is clearly impossible because the wind must retain some kinetic energy in order to move away from the turbine after passing through it. This limitation on energy extraction from the wind is known as the "Betz Limit" (Okuluv and Sorenson, 2008) or the "Lanchester–Betz–Joukowsky limit" (Sorenson, 2011). This limit on kinetic energy extractions means that the theoretical maximum amount of energy that can be extracted from wind moving through a turbine is 59.3% (Lam, 2006).

The Betz Limit can be demonstrated using aerodynamic theory, but there are also some other limitations on turbine performance. In practical situations wind turbines are affected by air drag and friction on rotor blades. The swirling effect of rotating turbine blades creates eddies and the air slows as it passes through the turbine, affecting the tubular representation of wind passing through a turbine. In practical turbines the tubular effect tends to expand (Sorenson, 2012).

The Power Coefficient (C_p) indicates how efficient wind turbines are in converting available wind energy into useful energy. This means that practical wind turbines will supply less energy than the 59% limit set by Betz.

To be useful, the turbine power must be delivered to consumers in the form of electricity. Large scale systems normally supply into an electrical grid system. Wind power reliability must be factored into the design of these connections. The electrical power generated by wind turbines is described as intermittent or variable because it depends on if, and when, wind blows. It is likely, therefore, that a building supplied with wind power generated electricity will also be connected to the supply authority grid from which power will be available if the wind stops blowing.

On a larger scale, wind farms are being developed in Europe and elsewhere that are designed to contribute energy to regional supply grids. Reliability of supply is important to supply authorities and is a statutory duty in the UK. Part of the strategy of ensuring reliability is to incorporate some reserve generation to cope with demand fluctuations, but the closer the available power matches the demand, the more efficient the overall system will be. Engineers must try to predict demand fluctuations as well as coping with the possibility of unexpected failure of some generating plant.

Despite the variable nature of wind, it is likely that wind will play a part, alongside conventional power stations, in supplying electricity to the grid and may reduce the amount of power that is necessary from conventional/thermal power stations. However, wind will not replace conventional power stations on a one-for-one basis. For the UK it is estimated that 8000 MW of wind could displace 3000 MW of a conventional (gas-fired) plant (Boyle, 2007).

6.5 Solar hot water

Despite the fact that the UK has a reputation for grey and cloudy weather, generating domestic hot water using solar energy in the UK is still considered to be a practical proposition. However, it is unlikely that a UK building could power 100% of its annual domestic hot water demand by solar energy alone.

Figure 6.6 Wind turbines
(© Macmillan New Zealand)

A solar hot water system operates by circulating water through a solar collector, which is typically roof-mounted. The system may be active or passive and it may also be direct or indirect. Active systems use a pump to circulate the water. A passive system is one in which the pressure required to circulate the hot water is obtained from the natural convection effects caused by heating water. It is known as "gravity" circulation or circulation by thermo-siphon. Different heat transfer mechanisms, adding natural convection, vapour boiling, cell nucleus boiling and film wise condensation are observed in the thermo-siphon solar water heater with various solar radiations (Hossain et al., 2011).

Direct systems are those in which the water that is heated in the solar collector is used as domestic hot water (Al-Nimr et al., 2011). An indirect system is one in which the water heated in the collector transfers its heat to the domestic hot water in a heat exchanger. The primary hot water in an indirect system never comes in contact with the hot water, which will be used for bathing and so on. UK practice tends to favour active, indirect systems for the following reasons (CIBSE, 2009a):

- Passive systems require the storage cylinder to be mounted above the collector. This is not practical in UK buildings.
- Active systems use a pump to create circulation. This offers greater control and can mean smaller pipework.
- Pumped primary hot water in an indirect system can contain an anti-freeze solution. Avoiding damage from frozen pipe systems is an important consideration in the UK.

Solar hot water systems in the UK will normally require a back-up heat source for those occasions when the available solar energy cannot heat the domestic hot water to the required temperature (Currie et al., 2008).

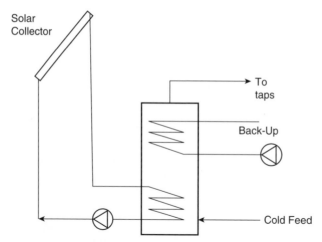

Figure 6.7 Solar hot water system with a back-up heat source
Adapted from Solar Heating Design and Installation Guide (2007) with the permission of CIBSE

In the UK domestic hot water in residential properties is normally heated in one of three ways. Hot water storage heated by electric immersion heaters was very common. It is now more likely that hot water demand is met by instantaneous gas-fired heaters that are also part of the building heating system. The other alternative is to heat stored water using a heating coil within the storage vessel. The heating coil receives heat energy from the central heating system.

The diagram in Figure 6.7 depicts a domestic hot water storage vessel that contains a coil heated by solar energy and a back-up coil from the building central heating system. The part of the system that absorbs the solar energy is called the collector. Roof-mounted solar collectors absorb solar energy to heat water, which is circulated through the coil in the storage vessel (CIBSE, 2009b). In order to be functional the solar collector must be mounted at a suitable angle, face the sun and not be over-shadowed by trees or other buildings (Camacho et al., 2012). Solar collectors are normally permanently mounted in a fixed location. Therefore a suitable angle for mounting is selected. Ideally, in the UK, the collector will be mounted at 30° facing directly south. Buildings do not always have directly south-facing roofs and solar collectors can provide useful energy at orientations between south-east and south-west. There are two main types of collectors:

- Flat plate
- Evacuated tube

A flat plate collector consists of a flat plate upon which is mounted copper or aluminium tubing. The plate is dark coloured to improve solar absorption. Flat plate collectors normally have a transparent cover. The collector is mounted within an insulated case.

Evacuated tube collectors use the same principle as vacuum flasks. An example of evacuated tube solar collector and hot water storage assembly is given in

Figure 6.8 Evacuated tube solar collector and hot water storage assembly

Figure 6.8. The absorber is set within a glass tube that is under vacuum pressure. The vacuum provides improved thermal insulation. Of course, seals and workmanship play an important part in their construction.

Not all of the solar energy that strikes a collector transfers heat into the system fluid. The efficiency of a solar collector describes how much of the solar radiation striking the collector is converted into useful heat energy. The efficiency of a solar collector is a key factor for the performance of thermal facilities. As the weather conditions vary continuously during the day, the instant collector efficiency depends not only on the components employed in its construction but also on the actual environmental conditions, the hot water temperature and aging (Rodriguez-Hidalgo et al., 2011). Some of the radiation will be reflected from the surface and some will be absorbed, as schematically shown in Figure 6.9. Some of the energy absorbed by the absorber plate will transfer into the water. Thermal losses from the collector will be in proportion to the temperature difference between the air temperature and the collector temperature. Higher ambient air temperatures will increase thermal losses. Direct and diffuse solar radiation is absorbed by the collector surface. The collector efficiency is affected by reflection and by thermal losses.

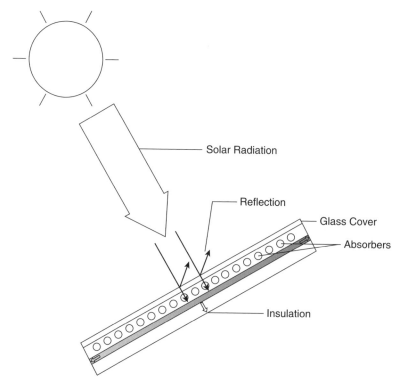

Figure 6.9 Reflection and absorption of solar radiation

6.6 Conclusion

This chapter has provided an insight into the use and development of low-carbon technologies in UK buildings. Obviously, the energy requirements of a building are greatly influenced by the particular climate of its location. However, growing infrastructure needs in much of the world require that technology will play a vital part in maintaining modern societies.

Modern lifestyles mean that man-made infrastructure will be a vital necessity and it is now recognised that construction professionals must play an important role in managing the infrastructure sustainably.

The next chapter focuses on sustainability in utilities and water-efficient sustainable buildings, setting out the broad principles of domestic water efficiency as providing an adequate, secure, domestic water supply for intensely urbanised, developed societies is now a growing problem worldwide.

References

Solar hot water

Al-Nimr, M.A., Khuwaileh, B. and Alata, M. (2011). A Novel Integrated Direct Absorption Self-Storage Solar Collector. *International Journal of Green Energy*, 8, 618–630.

Camacho, E.F., Berenguel, M., Rubio, F.R. and Martínez, D. (2012). Control of Solar Energy Systems. Berlin: Springer.

CIBSE (2007). Solar Heating Design and Installation Guide. The Chartered Institution of Building Services Engineers.

CIBSE (2009b). KS15 Capturing Solar Energy. The Chartered Institution of Building Services Engineers.

Currie, J.I., Garnier, C., Muneer, T., Grassie, T. and Henderson, D. (2008). Modelling Bulk Water Temperature in Integrated Collector Storage Systems, Building Services Engineering Research and Technology, 29, 203–218.

Hossain, M.S., Saidur, R., Fayaz, H., Rahim, N.A., Islam, M.R., Ahmed, J.U. and Rahman, M.M. (2011). Review on Solar Water Heater Collector and Thermal Energy Performance of Circulating Pipe, Renewable and Sustainable Energy Reviews, 15, 3801–3812.

Rodriguez-Hidalgo, M.C., Rodriguez-Aumente, P.A., Lecuona, A., Gutierrez-Urueta, G.L. and Ventas, R. (2011). Flat Plate Thermal Solar Collector Efficiency: Transient Behavior under Working Conditions. Part I: Model Description and Experimental Validation, Applied Thermal Engineering, 31, 2394–2404.

Energy efficiency

Bell, L. (2013). Total Sustainability in the Built Environment. Basingstoke: Palgrave Macmillan, 54–55.

CIBSE (2012). Guide F: Energy Efficiency in Buildings. The Chartered Institute of Building Services Engineers.

CIBSE (2014). Design for Future Climate Case Studies: TM55. The Chartered Institute of Building Services Engineers.

European Journal. (2003). The EU Energy Performance of Buildings Directive. (Available at http://publications.europa.eu/official/index_en.htm, accessed 2 July 2015).

Schobert, H. (2013). Energy: The Basics. Abingdon: Taylor and Francis, 163–225.

Zero Carbon Hub (2014). Zero Carbon Policy. (Available at http://www.zerocarbonhub.org/zero-carbon-policy/zero-carbon-policy, accessed 2 July 2015).

Wind turbines

Boyle, G. (2007). Renewable Electricity and the Grid: The Challenge of Variability. London, UK: Earthscan Publications Ltd.

Lam, G. (2006). Wind Energy Conversion Efficiency Limit. Wind Engineering, 30, 431–437.

Okuluv, V.L. and Sorenson, J.N. (2008). Refined Betz Limit for Rotors with a Finite Number of Blades. Wind Energy Journal, 11, 415–426.

Sorenson, J.N. (2011). Aerodynamic Aspects of Wind Energy Conversion. Wind Energy Journal, 43, 427–448.

Sorenson, J.N. (2012). Aerodynamic Analysis of Wind Turbines. Comprehensive Renewable Energy, Editor-in-Chief: S. Ali, ed., Elsevier, Oxford, pp. 225–241.

Recommended websites

European Commission, Energy Efficiency. (Available at http://ec.europa.eu/energy/efficiency/buildings/buildings_en.htm, accessed 2 July 2015).

The Carbon Trust. (Available at http://www.carbontrust.com/resources/reports/advice/conversion-factors, accessed 2 July 2015).

The Danish Wind Industry Association. (Available at http://www.windpower.org/en/, accessed 2 July 2015).

The Sector Skills Council for Building Services Engineering. (Available at http://www.summitskills.org.uk/renewables/376, accessed 2 July 2015).

UK Govt. Department for Communities and Local Government. (Available at http://www.communities.gov.uk/planningandbuilding/sustainability/energyperformance/, accessed 2 July 2015).

7 Sustainability in Utilities: Water Efficient Sustainable Buildings

David Phipps and Derek King

7.1 Introduction

Water scarcity and water stress are now affecting many parts of the world and a huge amount is being written about these topics, with contributions coming from a whole range of organisations. The United Nations, the European Union, national and local governments all contribute extensively, along with non-governmental organisations, research organisations, universities and charities who are also heavily involved. A selected bibliography of their output, together with other material, is provided at the end of this chapter.

About 1% of the Earth's water is freshwater available for human needs and much of this limited supply of freshwater is polluted and unsuitable for use without further extensive (and expensive!) treatment. Nevertheless, in the developed world, water has been treated as a limitless resource and people have become accustomed to having a continuous supply of high quality potable water (water that is safe for humans to drink) available on demand. Although in principle water is a "renewable resource" through the hydrological cycle, if it is used at a faster rate than it can be replaced the resource becomes depleted and will eventually no longer be available. Many parts of the world are already subject to some form of water stress evident in the deterioration of freshwater resources in terms of quantity (aquifer over-exploitation, dry rivers, lakes etc.) and quality (eutrophication, organic matter pollution, saline intrusion etc.). Highly developed urban societies cannot easily solve the problem of water stress; in fact just the opposite is true. Many of the world's major conurbations face increasing problems and have reached limits of supply. South-Western USA, many parts of Europe, south-Western Australia and of course urban megacities in India and China are among the worst affected. It is now widely acknowledged that water stress is a growing threat in the developed world. It is in this context that attention is being increasingly focused on the sustainable use of water.

A series of factors, often working together, are exacerbating the problem of water stress:

- the climate is changing[1] leading to changes in rainfall,[2] glacial melt and surface water evaporation with further adverse effects on freshwater supply;
- individual water demand is growing;
- increased agriculture and industry compete with domestic use for scarce water resources;
- the population is increasing and consequently aggregate demand for water will increase;

- the majority of the world's population is urbanised, demand is highly localised so supplies of water do not easily match demand;
- demand for water in the developed world is irrational in that high quality potable water is used for all purposes, where for many uses, lower quality water, especially from recycled sources, would be appropriate.

The result is that the current pattern of consumption is unsustainable. To manage this problem, supply and demand must be brought into balance. Increasing supply through collection or abstraction is becoming ever more difficult, if not impossible. If new sources of water are unavailable then more efficient use of existing supplies becomes ever more important. This can be divided into two phases. The first phase, primary conservation, brings about more efficient use of the core supply, the second phase is to capture and re-use the wastewater from the first use. However, wastewater re-use faces obstacles that include public reluctance, technical and economic problems and hygiene risks.

Demand control involves a complex inter-relationship between technical innovation (i.e. water efficient devices), economic policy (price and metering) and the habits and attitudes of the user (bathing vs shower, use of running water for cleaning teeth, watering the garden etc.). Although technology is important, public attitudes and personal behaviour regarding water use must change substantially if demand is to be controlled.

Although the primary purpose of water saving is seen as maintaining a supply, water saving is inextricably linked to energy saving. It is obvious that collecting and purifying and transmitting water is energy dependent. The use of hot water is particularly energy demanding and the subsequent collection, transmission and treatment of wastewater also requires energy so that any reduction in water use should be energy saving.

Like all schemes, water efficiency measures have thrown up a whole series of unintended and often adverse consequences. The reduction of water going to drain means that flow in the sewers is lower, which may mean that sewers are no longer effectively self-cleaning. The wastewater reaching the treatment works will be lower in flow and higher in organic matter, perhaps necessitating a re-configuration of the works or at least a change in operation.

One interesting development is the introduction of separate potable and non-potable water supplies. For example several cities in Utah, USA have installed separate "purple pipe" distribution systems to supply irrigation water separately from potable water. However, these are small communities and it is difficult to see how this might be applied to larger areas.

Designing water efficiency into new-build developments is now accepted practice. Numerous schemes have demonstrated different approaches and there are well developed regulations and supporting guidelines. With new builds, developers have not only the opportunity to use water efficient devices, such as low-flow showers, dual-flush or low-volume flush WCs and so on, as discussed later, but also to include schemes for rainwater harvesting and greywater[3] re-use. These can operate locally at the level of a single building, co-operatively for small communities or extensively for large buildings. Unless the development itself is large, projects such as stormwater retention and direct potable re-use are likely to be too costly and so are unlikely to be adopted unless driven by legislation or extreme water scarcity. In addition, there is a psychological barrier to consumer acceptance with a marked reluctance to accept that foul water can be sufficiently treated to make it safe to drink.

In most places building regulations and codes are being updated to demand better water efficiency. For example, in Australia, the Queensland Government introduced new mandatory water saving targets for applications lodged for construction of new houses in South East Queensland from 1 January 2007. Interestingly, water tanks, individual and communal rainwater tanks and storm-water re-use are included as one means of achieving the water savings target. The aim under the title *Target 200* was 200 litres per person per day, which is surprisingly high by standards elsewhere. England has a target of 125 litres per person per day, which has been applicable for all new-builds as set out in the various amendments to Part G of Building Regulations 2000 and related documents such as the Water Efficiency Calculation Methodology from the Department For Communities and Local Government. However, the pace of retrofit is less certain.

Commercial buildings, such as large office blocks, shopping malls and other larger scale developments may see more advanced water efficiency measures introduced once the economic case is made or when obliged by legislation to undertake such measures.

Retrofit to improve the efficiency of existing buildings is a different problem. The only feasible action on a large scale is to encourage consumers themselves or their landlords to fit and use water efficient devices, with consumers modifying their lifestyles as necessary. An extensive study by Waterwise on the water saving potential for social housing in London for the Greater London Authority and the effect of economic incentives by the London Climate Change Partnership have explored the problems of retrofit in showers very thoroughly and detailed recommendations are made. A key point, very clearly recognised in the study, is the sheer scale of the problem. Given the lifespan of many domestic fixtures and fittings this is likely to be a slow process unless accelerated by clear, coherent and acceptable initiatives, which individuals and households could adopt. The training and skills of plumbers would be important, for example registration schemes/competent person schemes as for gas and electricity.

Despite the prolonged timescale needed, municipal authorities are encouraging retrofit. For example in the USA, Tampa, Florida has subsidised a plumbing retrofit kit. A similar scheme has been considered for London with a subsidy for water efficient shower heads and toilets proposed, as mentioned previously. This study makes some key points about the size of subsidy required to effect significant uptake in any scheme. Financially, retrofit programmes can be judged on some form of cost-benefit analysis, though it may be difficult to obtain meaningful answers. Several schemes were compared as part of the decision making process for London. These included Calgary's toilet replacement rebate scheme, the showerhead replacement programme in York Region (Canada) and Sydney's washing machine replacement rebate programme. Superficially, Calgary with a benefit-cost ratio of 42.9[4] appears to be the most successful. This remarkable result was judged to be due to the combination of three factors:

- Calgary has the highest water supply cost;
- the high flushing volume of the toilets being replaced (19 l);
- the low value of the rebate given.

For example, subsidy programmes in Sydney and Austin, Texas found that a rebate of around 50% of the price difference was sufficient to entice consumers to purchase high efficiency washing machines. Excluding Calgary, the other programmes had benefit-cost ratios above 3, but for different reasons. York Region's programme was based on the promotion of basic free conservation devices. Retrofitting toilets was more expensive but replacing showerheads was relatively cheap though both had significant water saving benefits. In Sydney, the expensive washing machine rebate programme was judged to be a success as it was based on the economic incentive of the relatively high cost of supplying water in the city. In contrast the other two washing machine rebate programmes (Austin, Queensland) had a benefit-cost ratio of less than 1. All three washing machine programmes had similar water savings; but the low cost of supplying water in Austin and Queensland reduced their short-term economic effectiveness. However, when the longer term view was taken and the costs arising from water shortages in drought periods were avoided their programmes could be justified.

The question of how legislation might affect the pace and direction of retrofit is one that needs further attention. However, evidence from the introduction of energy efficiency measures shows that the uptake of efficient devices can be driven through education, albeit slowly, but for rapid progress and high levels of adoption mandatory inclusion in building regulations and codes is necessary.

7.2 Alternative sources of water for the sustainable building

7.2.1 Water reclamation and re-use

Water reclamation and re-use provides a means of augmenting traditional first-use water supplies. Water re-use can help to provide the recycle loop between water supply and wastewater disposal. If water can be recycled multiple times for various purposes, each less demanding on quality than the previous one, then the net water demand can be considerably reduced. Water re-use issues have been extensively considered but schemes are still relatively uncommon.

Like all recycling systems, the techniques described next are not a universal prospect. Whilst it may be easier to incorporate some schemes in new-build projects it is much more difficult to apply as a retrofit. Nevertheless, as water scarcity increases these types of schemes will become more attractive and even essential.

Effective and reliable water re-use requires the development of a new infrastructure either at individual homes (greywater) or system level (direct and indirect potable re-use). This means that the case must be made for the reliability of the treatment process with satisfactory economic and financial analyses. A key factor limiting the implementation of water re-use systems is often the cost of the infrastructure needed.

If the water re-use system is community based, with an upgraded wastewater-treatment process it may be technically or financially practical to return reclaimed water to urban areas to replace some primary potable water supply. The use of distributed water re-use systems with satellite and decentralised configurations may be helpful. However, even when the reclamation is technically and financially satisfactory, public acceptance cannot be assured and much work may have to be done to gain support.

7.2.2 Rainwater harvesting

Rainwater harvesting is attractive in its apparent simplicity. However, it is not always straightforward. First, the space to construct the storage needs to be available. There is a problem in that unless the storage is sufficient to deal with longer dry spells it will work best when least needed. Then the water quality will depend on the type of roof and capture conditions, especially the level of organic material that might be collected. In both the short and long storage term the growth of algae and other micro-organisms, which may be pathogens, needs to be considered. Other pollutants deposited on the roof may also have adverse effects. Nevertheless, rainwater harvesting from roofs can offer an additional water supply for individual households, single buildings and small groups of buildings. At a larger geographic scale, stormwater harvesting can augment the normal water supply for urban communities. However, like all sources of untreated and non-potable water, the quality of stored rainwater stored is very variable. Although widely used in domestic gardening, rainwater has really not found widespread use as a direct augmentation or replacement for treated mains water in developed countries.

7.2.3 Stormwater harvesting

Stormwater is unfiltrated run-off from impermeable areas such as roads, car parks and the like. Harvesting stormwater involves civil engineering to collect and store the water. It also requires the ability to monitor quality as stormwater carries pollutants and pathogens that need to be treated to a level appropriate to the use of the water. Stormwater harvesting requires a large investment in constructing and maintaining storage facilities and the economics of such projects are complicated to evaluate. Nevertheless, stormwater harvesting can contribute significantly to urban water supplies and many schemes are being investigated. For example, as part of the Australian National Water Initiative, schemes of stormwater harvesting are being investigated to replace demand on the potable water supply. Such schemes can have added value through flood mitigation, the provision of urban wildlife habitat and in some cases provide scenery for recreation value of urban parks. In general, the less secure the primary supply and the more variable the weather, the more important stormwater is likely to become.

7.2.4 Greywater

Greywater, or sullage, is the combined residues from household use, bathing, food preparation, dishwashing, cleaning, laundry and so on, excluding the waste containing faecal matter and urine, which is described as blackwater. If greywater re-use is intended, the routing of it into a collection systems unconnected to blackwater is undertaken. The use of greywater can be immediate, for example for irrigation, or deferred, for example for WC flush, which depends first on satisfactory collection and storage. Given the poor quality of all aspects of greywater, physical, chemical and microbiological (including viruses) storage should be minimised, thereby restricting the opportunity for further degradation. Until recently greywater re-use, other than informal irrigation, was discouraged or even illegal in many parts of the world. This situation is now rapidly changing and in many jurisdictions schemes of approval and licensing are being enacted.

Other considerations include deciding whether treatment is required or not and, if treatment is required, how sophisticated does it need to be? What maintenance is required and what are the capital and recurrent costs? If greywater is to be used, what will be the public perception and what will user acceptance have to be? This will depend on the types of use. For example, greywater used internally for WC flushing may need extra cleaning, if only for aesthetic purposes, compared with use in the garden irrigation. Garden use appears attractive as it can readily displace the use of potable water for irrigation. However, continued use of greywater can have a substantially negative impact and damage the soil by adding material that is not easily biodegradable. A growing body of research on the consequences of greywater use is appearing and as both the advantages and problems become more clearly understood, its use will continue to increase. Regulations and guidelines on re-use differ from country to country. *Article 12 of the Urban Wastewater Treatment Directive 114* indicates that treated wastewater shall be re-used whenever appropriate. Greywater treatment and re-use could produce significant water savings but without comprehensive removal of hazardous substances eco-labelling and regulatory controls will be required. It is interesting to see that packaged single dwelling greywater systems are commercially now available.

7.2.5 Blackwater

Blackwater is wastewater containing faecal matter and urine. It is also known as brown water, foul water or sewage. It is quite distinct in this respect from greywater and treatment is almost always required before further use. Since the treatment of this type of wastewater is complex and demanding, blackwater re-use is far less common than greywater recycling. However, blackwater has one major attraction; it carries the major portion of the total plant nutrient content (NPK etc.) in wastewater. It is therefore increasingly attractive to capture these if it can be done safely and economically. Phosphate recovery is particularly important given the rapidly diminishing supply for fertiliser use. Usually, however, because of the technologies required, recapture is normally only carried out at a large scale. In contrast, in Scandinavia, small scale systems have been developed and tested for rural areas without local area sewage system. The development of such systems could have major implications on the environment, public health and recycling of nutrients to the land.

7.2.6 Direct potable re-use

Not to be confused with direct potable use, which is the use of untreated sources such as wells, streams or even captured rainwater, direct potable re-use is the introduction of recycled water directly into a potable water distribution system, downstream of the wastewater treatment plant. Given the complex nature of waste water, and the difficulty of economically treating it to the required standard, this technique is practically very difficult to apply. There is no doubt that the public's strongly negative perception and attitude to this, particularly when first described to them, is a barrier to acceptance.

7.2.7 Indirect potable re-use

One way to tap into an unused water reserve is indirect potable re-use (IPR). IPR is a treatment process where highly treated municipal wastewater is discharged into an environmental buffer, that is, a natural freshwater source such as a river, lake or groundwater storage. Whilst greywater use specifically avoids health sensitive functions, IPR deliberately allows suitably treated water to re-enter the supply cycle. Whenever wastewater is reclaimed for human consumption, the treatment process must be reliable and capable of removing pathogens and other contaminants present in the water. Reverse osmosis (RO) is currently the only technique capable of removing the required pathogens and ultrafiltration (UF) has proven to be an excellent pre-treatment solution to remove suspended solids and bacteria and help maximise the RO performance.

Unplanned IPR has existed even before the introduction of reclaimed water. Many cities already use water from rivers that contain effluent discharged from upstream sewage treatment plants. There are many large towns on the River Thames upstream of London (Oxford, Reading, Swindon, Bracknell) that discharge their treated sewage into the river, which is used to supply London with water downstream. This places an additional burden on the treatment required before the water can be considered safe for human use.

7.3 The sustainable future: setting targets and embedding outcomes

7.3.1 Demand reduction

Demand reduction requires users to change their habits, usually linked to a technology change, for example water efficient domestic appliances. Recruitment to a modified lifestyle is easier than the long-term retention of those recruited. Studies on consumer adoption of water saving measures are still relatively few, though gradually increasing, but almost all deal mainly with factors affecting recruitment and retention is given less prominence. A significant problem is the authentication of measures at user level. It is well known that users will remove or by-pass water saving devices.

7.3.2 Where are we going? – Setting targets

Water consumption in the developed world varies widely but typically ranges from 150 to 520 litres per person per day for all uses combined. The estimate for minimum requirement is less than 10 litres per person per day (lpppd). The target for sustainable consumption varies, but in the UK a target of 130 lpppd (or less) is now established and there will be continued pressure to reduce this further. Similar targets have been set for Europe, the USA and for many other countries. The target reflects the estimate of what is achievable over the short to medium term. In the UK Building Regulations have been supplemented by the Code for Sustainable Homes. Similar activities are in progress in most other developed countries.

To ensure water supply into the future, cities must undergo a paradigm shift and become "water sensitive", reducing demand by more effective use, minimising waste and recycling. Water sensitive cities need an integrated urban water cycle management that takes account of natural hydrological and ecological cycles. This approach promotes co-ordinated planning for sustainable development via the management of water, land and related resources (including energy use) linked to urban areas and the application of water sensitive urban design principles.

It is readily possible to construct energy efficient buildings. Ultra-low energy buildings such as PASSIVHAUS (passive house), that require little or no input of generated energy for space heating or cooling, are well known. The Passivhaus standard is a rigorous, voluntary code for energy efficiency in a building, aimed at reducing its ecological footprint. A similar standard, MINERGIE-P, is used in Switzerland. In principle, a water-neutral house could also be constructed. However, water efficiency is not normally thought of in the same way as carbon, and it is accepted that water will be abstracted from the supply used and passed on. All water could be continuously recycled like a space-craft but the cost in terms of initial investment in equipment and the ongoing energy requirement would be prohibitive.

7.4 External controls on water use

7.4.1 Education, advice and legislation

To be effective, the movement towards water efficiency requires at least three strands – education to show the public that there is a problem, advice on voluntary water saving measures and legislation to enforce standards. Education and advice on water stress and water efficiency is widely available. Organisations such as Waterwise and Water UK, which represent all UK water and wastewater service suppliers at national and European level, provide a positive framework for the water industry to engage with government, regulators, stakeholder organisations and the public. All local authorities in the UK have some input into public education. Similar processes operate elsewhere.

7.4.2 Legislation, regulations and codes of practice

Legislation sets the framework for mandatory standards, which are then defined in regulations that are finally interpreted in codes of practice and guides. In general there needs to be a mix of regulation and persuasion. For example, in England, initiatives such as the Code for Sustainable Homes, or regulations such as the Water Supply Regulations 1999, are matched by governmental effort through the Environment Agency's Market Transformation Programme.

For England, the updated Part G of the Building Regulations came into effect in April 2010. The new draft sets a whole building standard of 125 litres per person per day for domestic buildings. This comprises internal water use of 120 litres per person per day, and in that respect is in line with Code Levels 1 and 2, plus an allowance of 5 litres per person per day for outdoor water use. This will be specified using the methodology set out in the "Water Efficiency Calculator for New Dwellings", also used for the Code for Sustainable Homes. An introductory guide for housebuilders provides assistance in understanding the requirements and means of implementation.

As examples of the drive to efficiency in the USA, The San Diego Municipal Code 147.04 requires that all residential, commercial and industrial buildings must be certified as having water-conserving plumbing fixtures in place prior to a change in property ownership. Similarly, the City of Manhattan requires all residential buildings to be retrofitted with a high efficiency toilet before the property can be sold. Burbank, California gives details of standards to be achieved for all new sales after 23 August 2010; both the seller and the buyer of the property will need to certify that the property meets or exceeds the specified water use standards.

There is a need for an independent measurable set of standards and benchmarks, as for example with BREEAM[5] (Building Research Establishment Environmental Assessment Method), an environmental assessment method and rating system for buildings from BRE global that replaced the previous EcoHomes standard. Similarly, LEED (Leadership in Energy and Environmental Design), developed by the U.S. Green Building Council promotes sustainable building and development practices through a suite of rating systems that recognise projects that implement strategies for better environmental and health performance. IAPMO (The International Association of Plumbing and Mechanical Officials) has published a "Green Plumbing and Mechanical Code Supplement", which serves to complement *any* adopted plumbing and mechanical code by bridging the disconnection between existing codes and established green building programs. The Green Supplement addresses areas such as:

- use of alternate water sources (greywater, rainwater harvesting);
- proper use of high-efficiency plumbing products;
- conservation of hot water;
- energy conservation in HVAC systems.

Another approach is to set up labelling schemes which allow the perceptive consumer to choose water efficient devices. The Australian Water Efficiency Labelling and Standards (WELS) Scheme is well known as a leader in this area. However, there is no universal standard nor is there always agreement on the approach to be taken between different advisory codes.

7.4.3 Different approaches – whole building vs point-of-use

New build offers significant opportunities, but not without careful consideration. For domestic dwellings, a whole building approach might be complex to implement but could be mandated and made to work. Performance could be set at a whole building standard, say in the range of 120–135 lpppd. If properly drawn up, a code could provide clear guidance and would offer flexibility to developers and property owners whilst delivering water savings without unduly restricting consumer preferences. In contrast, point-of-use standards apply to individual fittings or identifiable sub-systems where items such as low volume WCs, aerating taps or low-flow showers are considered separately.

7.4.4 Pricing and metering

One way to incentivise greater efficiency is to charge on the basis of use, usually by metering. The demand for water could be restricted by a sufficiently punitive pricing structure. However, even if desirable, attempts to achieve restriction by pricing are subject to serious constraints. Water is essential to life and a "reasonable" supply of water is indispensable to maintaining the accepted pattern of standards of living. Water is seen as a public good and therefore not lightly to be subjected to the laws of supply and demand.

Improving efficiency through metering appears to be an obvious conservation measure. Once metering is established then charging regimes can be altered. In England, water metering is still only at about 30% of households, with the majority still charged on property value, though this is gradually changing as all new builds have to be metered. Metering is much higher in mainland Europe and elsewhere. Any scheme to manage consumer demand needs a considerable degree of consumer education. Metering can be introduced into previously unmetered communities. For example, after a major exercise, New York now has compulsory 100% metering.

A new development is the roll-out of smart meters. Smart meters can monitor, measure and communicate a customer's electricity, gas and water usage data to utility providers at intervals as little as 15 minutes. They can operate individually, linking only the single user and to the supplier or as part of a wireless "mesh" network with data collected and integrated from neighbouring meters. The data can be communicated to the user either by reading the meter or by use of a suitable app on a smartphone. The process will be slow. For example, Thames Water estimates that it will take until 2030 to have 100% metering. The claimed advantage is a reduction in water use, though the cost will fall on the consumer and there are concerns about data security.

The commonest pricing mechanism, the rising block tariff (RBT), has price rising stepwise as use increases. A three-tier format that has a low-cost "essential use" block, a standard block and a premium price block for non-essential use is common. In general it is presumed that a low or even a zero tariff applied to the first block can enhance affordability. For example, some base use of water may be free of charge with a steep increase above this level. Higher tiers are more expensive, thus creating disincentives for overuse. In principle this enables utilities to match revenues with the cost, to create a sustainable financial model, while at the same time providing subsidised water for basic needs at or below the cost of operations and maintenance. Rising block tariffs are designed to achieve several public policy goals, though there is considerable disagreement as to how effective they are.

It is claimed that this differential charging approach, based on consumption, allows households to get what they need, cheaply, and that they can then decide whether to consume more water in the expensive tariff block. However, the system is crude and in its basic form discriminates against those with less ability to reduce water use – those with larger household size, medical needs, susceptibility to heat stress or without resources to invest in water-efficiency to replace the existing domestic devices. Other factors may be a deterrent and are often outside the reasonable control of the users. For example, tenants are restricted in how they can alter appliances. In 2011, South West Water in England decided not to pursue the development of rising block tariffs as a means of curbing customer water use after a trial of 1000 homes had failed to produce significant change in the pattern of usage and so could not be used to justify the major changes required for a universal implementation of the scheme. Similar inconclusive trials have been reported elsewhere.

In a more detailed study by Wessex Water, using a combination of metering, rising block and seasonal tariffs applied to 6000 new occupants, it was suggested that overall water usage could be cut by almost one quarter. However, metering accounted for over two-thirds of this reduction and only a further 6% reduction was seen among customers charged on rising block and simple seasonal tariffs. The company noted that the overall results were encouraging but the imposition of seasonal and rising block tariffs were not an immediate prospect.

7.4.5 Social considerations – water poverty

Water poverty occurs when the cost of water is an excessive part of the household budget, even where water is available. Differential water pricing may therefore cause new affordability problems for households that are unable to reduce their water use. Rising block tariffs are sometimes coupled to subsidies. Providing a sub-sidised social tier is an effective strategy for reaching low-income households, but only if they are connected. Attempts to cross-subsidise from high-consumption (high-income) to low-consumption (low-income) households are effective only if a sufficient number of customers use the higher blocks. An obvious danger is that excessively high prices will drive users to alternate sources of provision.

Block tariffs can create other unintended disadvantages for the poor. For example, in developing countries households without private connections typically purchase water in relatively small quantities from private standpipe operators, water vendors and truckers. Since these vendors are bulk users they are re-selling water they have bought from the utilities at the highest block tariff cost. Similarly, when poor households group together to share a metered connection their aggregate consumption level pushes them into the higher price tiers. Water poverty is the lack of financial resources to obtain water. This is often a key determinant in the water economy at an individual level. There are conceptual problems in the definition and role of water poverty indicators and in the consequent policy issues: for this chapter the discussion will centre on the idea of water efficiency and sustainability in a modern, developed and largely urbanised society, where almost universal access to apparently unlimited potable water has heretofore been the norm. However, even where water supplies may be adequate not everybody benefits equally and water poverty is a growing problem. The Joseph Rowntree Foundation has defined water poverty as a household spending 3% or more of their income on their water and has warned that unless climate change is dealt with urgently, "water poverty" will become a serious problem in the UK for many households. It estimates that four million households in the UK are already "water poor". So, water efficiency is an important factor for everybody but particularly in the economy of the poorer members of developed societies.

7.5 Key targets for technical objectives

7.5.1 Basic plumbing – design, installation and maintenance

Although there is a lot of attention paid to water saving devices, the core of a well-designed and maintained systems is essential. A single dripping tap can waste over 5 m^3 of water annually. Leak detection at local level is possible using modern equipment but this seems unlikely to be widely adopted in the domestic market. However, water utilities are increasingly using sophisticated leak detection systems with advanced software to pinpoint hidden leaks.

The design of the system is important, though detailed specifications are usually left to the installer. Long pipe runs to hot water draw off points are a typical cause of excess water use as the water is run to clear the cold water in the pipe until a suitable temperature is reached. There are many national regulations, such as the American National Standard, the Uniform Plumbing Code (UPC) developed by the International Association of Plumbing and Mechanical Officials (IAPMO) that is updated every three years based on an open consensus protocol. In the UK, part G of the Building Regulations lays down standards for hygiene via the provision of sanitary and washing facilities, bathrooms and hot water. Part H of the same regulations deals with drainage and waste disposal. Similar regulations and codes are applied in most countries, for example The Australian National Construction Code (NCC) Series and many others.

7.5.2 WC

Typically, the WC uses about 30–40% of domestic water use and up to 90% for offices and public buildings. There has been a progressive reduction in the allowed volume of the flush. UK WC cistern volumes were 13 l until the 1960s when they were first reduced to 11 l and then further to 9.5 l in 1963. Since January 2001, the Water Regulations (1999), which replaced the previous Water Byelaws, stipulate that all new toilets installed in the UK have to have a maximum flush of 6 l, measured with the water supply turned off. Realistically, flush volumes will usually be higher than this, as water enters the cistern during the flush. It is generally accepted that with good pan design, full flush volumes down to 4 l do not present a problem in terms of flow for self-cleansing of drain liquid wastes.

Simple modifications of older existing WCs include devices such as putting a brick in the cistern or the slightly more sophisticated "Water Hippo", which simply reduce the volume available for the flush. This provides a very simple and cheap way of dealing with larger volume WC cisterns without having to fit a new system.

In the USA pre-1994 residential and pre-1997 commercial flush toilets use 3.4 US gallons (13 l) of water per flush (gpf or lpf). Since 1994 common flush toilets use only 1.6 US gallons (6.1 l). The trend to low flush continues. Before 1982, all WCs in Australia were single flush. The average flush volume of these toilets was about 11 l. Dual flush toilets were initially introduced with an 11/5-l configuration. In 1989 the 9/4.5 l model was introduced. After extensive trials involving retrofitting, a number of communities made these 9/4.5-l toilets mandatory. In 1995, after further extensive real-life testing, the volume was lowered further to 6/3 l. Then in 2005, the 4.5/3-l class was introduced. Attempts are being made to lower this further. In Scandinavia, 3/2-l dual flush toilets have been in use for some time.

There are problems with low volume flushes. In some cases, the solid faecal material is not fully removed. Improving flushing in conventional WC pans is one possible solution. Pressure assisted toilets claim to overcome this problem. They look identical to gravity toilets. The tank or cistern does not hold water directly but uses a sealed inner tank. When water enters the air inside the tank gets compressed. When the toilet is flushed, the water is expelled under the pressure of the compressed air which removes all the waste from the bowl much more efficiently than the water from gravity toilets. However, the device is more complicated than simple gravity systems. The Flushometer toilet uses a flush valve to control water direct from the main with pressures above 30 pounds per square inch (2.1 bar). They have no tank; hence they have zero recharge time, and can be used immediately by the next user of the toilet. This system requires larger diameter pipework to deliver a high volume of water in a very short time. The flush is very effective but this system still requires approximately the same amount of water as a gravity system to operate (1.6 gallons or 6 litres per flush) and is generally only found in commercial and public buildings.

Externally, the reduced self-cleansing in the drainage system may be a problem when very low volumes are used. For example, despite considerable objections, the San Francisco Public Utilities Commission has proposed using sodium hypochlorite to remove the odour and improve flow. Some ideas, such as WC with integral overflows, can create problems. In this case the overflow is easy to ignore and so water wastage can be increased.

Waterless composting toilets are not new. They rely on natural biological action to convert the toilet waste to useful compost. They provide an alternative to cess-pits and are well suited to low density housing which is not connected to main sewerage. However, the application of composting toilets to modern urban dwellings is not as simple as it first seems. To work properly dry, carbon-rich material is added to cover the fresh waste after each use. The compost needs monitoring to maintain the correct moisture level. Odour may be a problem and often they require an extractor fan.

7.5.3 The urinal

Waterless urinals are increasingly available in the mainstream bathroom fittings market. Although most private houses only have the WC and not a separate urinal, commercial buildings are increasingly turning to this technology. There are three principle types of waterless urinal, namely: microbiological, barrier and valve systems. Microbiological systems use micro-organisms to digest the urine contents so that no odour is produced either by a liquid seal in the barrier system or by a valve that traps the foul waste. For barrier and valve systems the flow of urine passes to the drain and odour is prevented from returning to the room's air, either by the liquid seal in the barrier or by the valve. There are advantages and disadvantages with each system. The commonest difficulty with all of them is that they require more maintenance than a simple water flush urinal. The bowl still needs cleaning and it is prone to block with hair and other debris. Some systems use a replaceable filter to capture the debris but this adds to running costs.

7.5.4 Source separation

The bidet and taps

Like most domestic fittings, taps are as much a fashion statement as a device for controlling water flow. There are, however, some technical advances that offer water saving possibilities. Aerating taps, which introduce air bubbles into the water thereby increasing its volume to give the same apparent flow, offer reductions of up to 50%. More for public areas than homes, taps with infrared sensors that provide water only once an object is underneath have been estimated to save up to 70% of overall use. These offer an alternative to mechanical devices, which have an auto shut-off after a timed length of flow. Taps with thermostats can keep the selected temperature, again offering savings of both water and energy.

Baths and showers

The volume of water in the bath can be reduced by lowering the overflow, which allows the bath to look large. The reduced depth, however, may be dissatisfying for some users. Alternatively, shaping the bath can reduce the overall volume while retaining sufficient depth for total immersion. However, this reduces the dimensions and so makes the bath look small. Baths are a fashion item as much as a utility device and the trend is still to have a standard size, especially in the replacement market.

Most websites, and other sources of water saving advice, regularly state that using a shower saves water when compared to using a bath. This is often, but certainly not always, true. As ever, the argument is a complex interplay between technology and human behaviour. The trite argument always presented is that a bath uses more water – the bath 130 l of water, a shower 5 minutes at 15 l per minute = 75 l of water – true so far. Even better with a water efficient shower giving, say, 6 l per minute flow. But what about a power shower? If this gives at 20 l per minute for 10 minutes, then usage is 200 l of water. Again what about showering frequency? Data is scarce but anecdotal evidence and some studies have shown that for certain groups, for example under 25s, showering may happen 3–5 times per day. Shower duration is also reported to be longer for this group. It is unlikely that they would take five baths a day. The point of this argument is not to disagree with the "shower better than bath" approach, but to say as always that simple mechanistic calculations may not present the whole truth. What is required is an approach that recognises that the whole system, both technology and user, need to be considered together.

Directly heated electric showers (tankless heaters), which are only connected to the cold water supply, can be installed on any type of system, provided there is enough electricity available. In the UK this is true but most countries in continental Europe have power supplies that are insufficient for this purpose. The available flow from direct electrically heated showers is inherently limited by the power available and the incoming water temperature. Taking some nominal UK winter conditions with incoming water at 5°C, the required temperature being about 45°C and the UK power limit for a normal 40 A domestic circuit as approximately 10 kW (240 V × 40 A), the maximum flow is about 3.6 l per minute. Clearly, if the incoming water temperature is higher and a lower water temperature is chosen, then a higher flow rate can be achieved. For example, in summer conditions with an incoming water temperature of 15°C and a target for the shower of 40°C a flow of about 6 l per minute is possible.

When just the transfer of electrical power to the water is considered, the direct electrically heated shower (tankless heater) is generally very energy efficient. In new equipment, efficiencies of >98% are achieved. This is far better than gas heating, which at best seems to give about 86% thermal efficiency. In both cases efficiency decreases with age, particularly in hard water areas, where scaling on the heater exchanger reduces heat transfer efficiency and also impedes flow. However, the argument is more complex since electricity derived from natural gas shows about 60% thermal efficiency.

Mixer showers, which need both hot and cold supplies, can be installed on any type of boiler system; pumped showers, however, can only be installed on storage-fed systems.

One strategy for reducing the shower flow rate without apparently reducing the "shower experience" is to use some form of aerating shower head in which the water is intimately mixted with a stream of air before leaving the shower head. Most large manufacturers have one or more such devices in their range and there are a number of patented devices that aim to use this approach. The design of the showerhead has become increasingly sophisticated with tools such as CFD (computational fluid dynamics) used to model the design and predict the performance.

The principal problem in persuading users to adopt a water efficient bathing habit is their perception of comfort, efficiency and their attitudes or personal preferences. Thus end users can be classified in two ways: the first based on awareness of water efficient habits and the second on the likelihood of them adopting water saving. Reports on home trials of showers where the overall flow is reduced generally show a marked reluctance by users to adopt low flow. This is in many ways surprising, as in the UK a large part of the market is taken by directly heated (tankless) electric showers that are unlikely to give flows much beyond 6 l per minute. Flow reduction for mixer showers, which can give flows well over 10 l per minute, can be achieved by using an easily fitted, simple restrictor, which reduces the diameter of the inlet pipe at its point of the attachment to the shower head. More complex devices offer a non-linear performance curve with flow allowed up to a set limit. These may be preferred as they do not interfere with shower performance at flows below the limit.

Clothes and dish washing

The use of a dishwasher is often criticised as being both water and energy inefficient compared to washing dishes by hand. However, as with showers, the argument is more complex and the outcome not so simple. It is now recognised that comparisons of different methods of dish washing are as much controlled by the behaviour of the individual as anything else. Washing dishes and rinsing in running hot water is wasteful both of water and energy. An average of about 10 lpppd is used when washing the normal amount of dishes by hand, whereas the best dishwashers use only just over 2 l per person to wash the same amount of dishes. The "average" hand-washing figures are debatable and depend on the actual washing regime. Equally, a full cycle in the dishwasher will consume about 15 l so unless the machine is fully loaded, the use per person per day will increase accordingly. Again, personal behaviour affects the usage – people still rinse dishes before placing them in the dishwasher, but this isn't required for modern dishwashers. In practice increasing dishwasher efficiency will not produce a huge saving per dwelling, as water used this way accounts for less than 2% of domestic use. However, as dishwashers are increasingly common, the aggregate saving could still be substantial.

Depending on size, top-loading washing machines use up to 150 l of water. Front-loading washers use far less water, normally between 50 and 100 l, with more modern machines generally using least. Using less water also gives a substantial energy saving as almost 95% of the electricity consumed by washing machines is used to heat the water. However, the main factor is the move to low temperature washes with modern detergents. Technical factors can only offer part of the solution as individual behaviour can substantially affect the frequency of use.

Gardens, car washing and swimming pools/jacuzzis

Activities such as car washing are gross users of water and are usually the first to be restricted when droughts occur. The use of greywater or rainwater to replace potable mains water has already been considered. However, other changes, such as using drought resistant plants and covering pools when not in use, are individual efforts that arise from either ecological awareness or economics and regulation.

7.6 The role of the consumer

Consumers can be described as advocates, allies, neutrals and opponents on the basis of their lifestyle and conservation behaviour. Advocates are committed to water saving and are prepared to spend time and money where available to reduce their water footprint. They will change their habits and also actively encourage others to do the same. Allies will moderate their own use, undertake normal maintenance and use water saving devices when updating their homes. Allies will make some adjustments to their lifestyle but purely on an *ad hoc* basis. Neutrals will either be unaware of water saving or disinterested. They are unlikely to persist with any lifestyle changes. Opponents will actively disregard water saving measures and persist in being water intensive. They will be careless in personal habits of water use or be high consumers, owning and using domestic irrigation systems, swimming pools, golf courses and the like in areas of water stress or shortage. They are the archetypical users of the golf course in an arid climate!

Consumer attitudes can be changed. Examples of successful campaigns are the introduction of smoking restrictions and the use of seat belts. Almost everywhere now there are governmental initiatives to raise awareness and implement water saving measures. One technique is to apply "choice editing", the process of eliminating "undesirable" choices, so that decision makers (of whatever kind) are still able to make choices, but only from a pre-determined sub-set. For example, it would be possible to ensure that the only showerheads on the market were low-flow models, though with sufficient variety to allow an element of consumer choice.

There is considerable tension between the commercial imperatives in the consumer lifestyle and the drive towards water efficiency, in many cases the change in social habits, particularly with regard to washing. Looking at just one sector, there is a huge effort towards water efficiency in showers, as discussed next, yet the marketing of power showers with flows of 20 l per minute or more, perhaps three times the acceptable maximum, is still widespread. This sector of the bathroom market is driven as much by fashion as by performance. The marketing of waterproof televisions for use in shower enclosures points to a real tension between efficient use of water for cleanliness and the luxury abuse of water for personal pleasure and satisfaction.

The recognition of the need to constrain and control the use of water has led to a wide variety of commercial opportunities that have been exploited with varying degrees of probity. The market place for bathroom fittings is extremely competitive but is generally perceived to be based on fashion rather than technology. Nevertheless the major manufacturers invest considerable resources in technical design. The appearance of aerating showerheads represents an attempt by manufacturers to satisfy water efficiency whilst at the same time gaining market share through innovation. Although manufacturers and retailers will give some information about flow rates and shower types, the principal selling features are usually cosmetic; appearance is a key distinguishing feature. Nevertheless, attitudes are changing and in 2012 the British Bathroom Manufacturers Association agreed an eco-labelling scheme similar to that already in existence for energy rating. Other eco-labelling schemes, such as the European Eco-Label, consider water saving as an integrated part of the life-cycle impact.

7.7 Conclusion

The water shortage experienced by many parts of the developed world will not easily be solved. Demand will increase, driven by urbanisation and population growth. It is difficult to see how infrastructure capacity will keep pace with urbanisation in developing countries. In developed countries, used to a permanent, accessible potable water supply, a real change in consumer attitudes and actions will become necessary. To achieve a real improvement in water efficiency a combined technological, educational, regulatory and financial approach will be needed. Perhaps the most effective influence would be water rationing, which would undoubtedly change the outlook.

A major problem is the timescale required for change. Without legislation and some form of compulsion water efficiency up-take will be slow. Bathrooms and kitchens can remain unchanged for decades and white goods (dishwashers and washing machines) may last many years. The pay-back for personal investment in water efficiency is poor unless water prices increase substantially. An important driver will be to link water saving, particularly for hot water, to energy saving and hence to apply a much higher rate of financial return which should greatly encourage up-take.

A water-sensitive dwelling should be metered, if only to monitor usage and not charge. The design will offer improved technical quality of water efficient fittings and white goods, which are independently assessed and eco-labelled. These will be installed and maintained according to an agreed code using approved inspected installers. Subject to site and practicality, water capture and water re-use for non-potable purposes will be designed within. Externally gardens and recreational spaces will be designed as necessary with plantings able to withstand drought. However, the key will be to sensitise the occupants and encourage them to adopt and maintain a water-efficient lifestyle despite the seduction of consumer pressure.

The next chapter deals with the legal aspects of the sustainable built environment and environmental impact assessment. The chapter emphasises the importance of international collaboration (i.e. policies, reports, protocols, conferences, laws, regulations and directives) and local actions for sustainable development.

Notes

1   Evidence is clear that there are major changes in the Earth's climate. The overwhelming consensus favours a global warming induced by increased release of "greenhouse gases", but, however they are being caused, climate changes will impose severe adverse effects on freshwater supply. In 2012 the end of nine years' drought in Australia was declared.
2   Note that drought is not the only problem, flooding is as serious a threat to water supply.
3   Greywater is used household water, excluding the waste containing faecal matter and urine which is described as blackwater.
4   The cost-benefit ratio attempts to summarise the overall value for money of a project. However, its full implementation may be complex.
5   For further information on BREAM and LEED see Chapter 5.

References

Attitudes and perceptions

Adams, D., Allen, D., Borisova, T., Boellstorff, D.E., Smolen, M.D. and Mahler, R.L. (2013). The Influence of Water Attitudes, Perceptions and Learning Preferences on Water-Conserving Actions. Natural Sciences Education, 42, 2013. (Available at https://www.agronomy.org/publications/nse/pdfs/42/1/114, accessed 2 July 2015).

Christensen, J. and Kowalski, M. (2006). Using Water Wisely: Quantitative Research to Determine Consumers' Attitudes to Water Use and Water Conservation. CCWater. (Available at http://www.ccwater.org.uk/wp-content/uploads/2013/12/Using-Water-Wisely-public-attitudes-towards-water-use-and-conservation-quantitative-research-CCWater-and-WRc-October-2006.pdf, accessed 15 July 2015).

Dolnicar, S. and Hurlimann, A. (2009). Drinking Water from Alternative Water Sources: Differences in Beliefs, Social Norms and Factors of Perceived Behavioural Control across Eight Australian Locations. Water Science & Technology, 60(6), 1433–1444. (Available at http://ro.uow.edu.au/cgi/viewcontent.cgi?article=1672&context=commpapers, accessed 2 July 2015).

Dolnicar, S. and Schäfer, A.I. (2006). Public Perception of Desalinated Versus Recycled Water in Australia. CD Proceedings of the AWWA Desalination Symposium 2006. (Available at http://ro.uow.edu.au/cgi/viewcontent.cgi?article=1145&context=commpapers, accessed 2 July 2015).

Murray, L. and Myant, K. (2006). Valuing the Water Environment: An Investigation of Environmental Attitudes and Values to Inform Implementation of the EC Water Framework Directive. Scottish Executive Social Research. (Available at http://www.scotland.gov.uk/Publications/2007/03/22131849/0, accessed 2 July 2015).

Climate change

Calow, R., Bonsor, H., Jones, L., O'Meally, S., MacDonald, A., and Kaur, N. (2011). Climate Change, Water Resources and WASH Working Paper 337 Overseas Development Institute, London. A Scoping Study. (Available at http://www.odi.org.uk/resources/docs/7322.pdf, accessed 2 July 2015).

WaterAid (2007). Climate Change and Water Resources. (Available at http://www.wateraid.org/documents/climate_change_and_water_resources_1.pdf, accessed 2 July 2015).

Codes and guides

Australian Building Codes Board (Watermark) (2014). (Available at http://abcb.gov.au/en/product-certification/watermark.aspx, accessed 2 July 2015).

Australian Capital Territory (2011). Water Efficiency Labelling and Standards Act 2005 A2005-10. (Available at https://www.comlaw.gov.au/Details/C2005A00004, accessed 14 July 2015).

Australian Government (2011). Water Efficiency Labelling and Standards (WELS) Scheme Strategic Plan 2012 to 2015. (Available at http://www.waterrating.gov.au/resource/water-efficiency-labelling-and-standards-wels-scheme-strategic-plan-2012-2015, accessed 14 July 2015).

BREEAM (2014). Website: (Available at http://www.breeam.org/, accessed 2 July 2015).

Department for Communities and Local Government (2006). Code for Sustainable Homes – A Step-Change in Sustainable Home Building Practice. (Available at http://www.planningportal.gov.uk/uploads/code_for_sust_homes.pdf, accessed 2 July 2015).

Department for Communities and Local Government. (2009). New Minimum Water Efficiency Requirements for New Buildings. (Available at https://www.gov.uk/government/uploads/system/uploads/attachment_data/file/7689/1234622.pdf, accessed 2 July 2015).

Department for Communities and Local Government. (2010). Code for Sustainable Homes Technical Guide. RIBA Publishing. ISBN: 978 1 85946 331 4. (Available at http://www.planningportal.gov.uk/uploads/code_for_sustainable_homes_techguide.pdf, accessed 2 July 2015).

Grant, N. and Thornton, J. (2009a). AECB Water Standards Delivering Buildings with Excellent Water and Energy Performance. Volume 1: The Water Standards. (Available at http://www.aecb.net/wp-content/uploads/2013/02/1503_AECB_Water_Vol_1_V3.pdf, accessed 14 July 2015).

Grant, N. and Thornton, J. (2009b) AECB Water Standards Delivering Buildings with Excellent Water and Energy Performance. Volume 2: The Water Standards. Technical Background Report. (Available at http://www.aecb.net/wp-content/uploads/2013/02/The_AECB_Water_Vol_2_V3.pdf, accessed 14 July 2015).

HM Government (1999). The Water Supply (Water Fittings) Regulations 1999 No. 1148. (Available at http://dwi.defra.gov.uk/stakeholders/legislation/ws(fittings)regs1999.pdf, accessed 2 July 2015).

HM Government (2000). The Building Regulations 2000: Sanitation, Hot Water Safety and Water Efficiency. (Available at http://www.planningportal. gov.uk/uploads/br/100312_app_doc_G_2010.pdf, accessed 2 July 2015).

IAPMO (2015). The Uniform Plumbing Code. (Available via purchase at http:// codes.iapmo.org/home.aspx?code=UPC, accessed 14 July 2015).

International Code Council (2011). 2012 International Plumbing Code (IPC). ISBN: 9781609830533.

LEED (2014). Website: (Available at http://www.leed.net, accessed 2 July 2015).

Queensland Plumbing and Wastewater Code. (2011). (Available at http://www. dlgp.qld.gov.au/resources/laws/plumbing/current/queensland-plumbing-and-wastewater-code.pdf, accessed 2 July 2015).

San Diego (1997). (Available at http://docs.sandiego.gov/municode/MuniCode Chapter14/Ch14Art07Division04.pdf and (Available at http://www.sand-iego.gov/water/pdf/wcc.pdf, accessed 2 July 2015).

The Australian Building Codes Board (Available at http://www.abcb.gov.au/en, accessed 2 July 2015).

Townsend, M. (2014) The International Code for a Sustainable Built Environment. BREEAM. (Available at http://www.breeam.org/filelibrary/ Technical%20Manuals/Code_for_a_Sustainable_Built_Environment_-_ BREEAM_Standards_for_Europe.pdf, accessed 2 July 2015).

Direct potable use

Crook, J. (2010) Regulatory Aspects of Direct Potable Reuse in California. National Water Research Institute NWRI-2010-01. (Available at http://nwri-usa.org/pdfs/NWRIPaperDirectPotableReuse2010.pdf, accessed 15 July 2015).

National Water Research Institute (2014). Direct Potable Reuse. (Available at http://nwri-usa.org/directpotable.htm, accessed 2 July 2015).

Tchobanoglous, G. (2013). Trends in Indirect and Direct Potable Reuse in the United States. (Available at http://www.watereuse.org/sites/default/files/ u8/California%20WateReuse%20GT.pdf, accessed 15 July 2015).

Greywater

Zita, L.T., Yu, Stenstrom, M.K. and Cohen, Y. (2014). Cost-Benefit Analysis of Onsite Greywater Recycling – A Case Study: The City of Los Angeles. (Available at http://innovation.luskin.ucla.edu/sites/default/files/Cost-Benefit%20 Analysis%20of%20Onsite%20Residential%20Graywater%20Recycling.pdf, accessed 2 July 2015).

Metering

Garner, V., Timlett, R. and Russell, N. (2011). Fairness on Tap: Making the Case for Water Metering. (Available at http://assets.wwf.org.uk/downloads/fair-ness_on_tap.pdf, accessed 2 July 2015).

Marchment Hill Consulting Pty (Swan forum). (2010). Smart Water Metering Cost Benefit Study. (Available at http://www.swan-forum.com/uploads/5/7/4/3/5743901/smart_metering_cost_benefit.pdf, accessed 2 July 2015).

OFWAT (2014). WaterSure (Vulnerable Groups Scheme). (Available at http://www.ofwat.gov.uk/consumerissues/assistance/watersure/, accessed 2 July 2015).

Thames Water (2014). Review of the Potential Health Effects of Smart Water Meters. (Available at http://www.thameswater.co.uk/Metering_website_-_Health_risks_of_smart_water_meters.pdf, accessed 2 July 2015).

Walker, A. (2009). The Independent Review of Charging for Household Water and Sewerage Services. DEFRA PB13336. (Available at http://www.defra.gov.uk/publications/files/walker-review-final-report.pdf, accessed 2 July 2015).

Rainwater harvesting

Environment Agency (2010). Harvesting Rainwater for Domestic Uses: An Information Guide GEHO1110BTEN-E-E. (Available at http://webarchive.nationalarchives.gov.uk/20140328084622/http://cdn.environment-agency.gov.uk/geho1110bten-e-e.pdf accessed 2 July 2015).

Re-use

de Koning, J., Miska, V. and Ravazini, A. (2006). Integrated Concepts for Re-use of Upgraded Wastewater: Water Treatment Options in Reuse Systems. AQUAREC EVK1-CT-2002-00130. (Available at http://www.amk.rwth-aachen.de/fileadmin/files/Forschung/Aquarec/D17-Aquarec-FINAL.pdf, accessed 2 July 2015).

Department of Water and Energy, State of NSW. (2008). Grey Water Fact Sheets 1–5; DWE 08_092 ISSN 1835-1360. (Available at www.waterforlife.nsw.gov.au, accessed 2 July 2015).

Environment Agency. (2011). Greywater for Domestic Users: An Information Guide. GEHO0511BTWC-E-E. (Available at http://www.sswm.info/sites/default/files/reference_attachments/ENVIRONMENT%20AGENCY%202011%20Greywater%20for%20Domestic%20Users.pdf, accessed 15 July 2015).

EPA. (2004). Guidelines for Water Re-Use 625R04108 (2004). (Available at http://water.epa.gov/aboutow/owm/upload/Water-Reuse-Guidelines-625r04108.pdf, accessed 15 July 2015).

The Council of the European Communities (1991). Urban Wastewater Treatment Directive (91/271/EEC). (Available at http://eurlex.europa.eu/LexUriServ/LexUriServ.do?uri=OJ:L:1991:135:0040:0052:EN:PDF, accessed 2 July 2015).

Urkiaga, A., de las Fuentes, L., Bis, B., Hernández, F., Koksis, T., Balasz, B. and Chiru, E. (2006). Handbook for Feasibility Studies for Water Re-Use Systems. EU Legal Deposit: BI-499-06. (Available at http://www.amk.rwth-aachen.de/fileadmin/files/Forschung/Aquarec/WP4-Aquarec-FINAL.pdf, accessed 2 July 2015).

Reviews

HM Government (2011). Water for Life. The Stationery Office Limited. ISBN 9780101823029. (Available at http://www.official-documents.gov.uk/document/cm82/8230/8230.pdf, accessed 2 July 2015).

UNESCO (2012). The UN World Water Development Report (WWDR4), 4th edition (3 Volumes) ISBN 978-92-3-104235-5. (Available at http://www.unesco.org/new/en/natural-sciences/environment/water/wwap/wwdr/wwdr4-2012/, accessed 2 July 2015).

UNESCO (2014a). The UN World Water Development Report Volume 1 (Water and Energy). ISBN 978-92-3-104259-1. ePub ISBN 978-92-3-904259-3. (Available at http://unesdoc.unesco.org/images/0022/002257/225741E.pdf, accessed 2 July 2015).

UNESCO (2014b). The UN World Water Development Report Volume 2 (Facing the Challenges). ISBN 978-92-3-104259-1. ePub ISBN 978-92-3-904259-3. (Available at http://unesdoc.unesco.org/images/0022/002257/225741E.pdf#page=153, accessed 2 July 2015).

Scarcity and quality

Abbott, J. (2011). Water Scarcity and Land Use Planning RICS, ISBN 978-1-84219-722. (Available at http://www.joinricsineurope.eu/uploads/files/WaterScarcityandLandUsePlanning.pdf, accessed 15 July 2015).

Brown, A. and Matlock, M.D. (2011) A review of Water Scarcity Indices and Methodologies. The Sustainability Consortium White Paper #106 (Available at http://www.sustainabilityconsortium.org/wp-content/themes/sustainability/assets/pdf/whitepapers/2011_Brown_Matlock_Water-Availability-Assessment-Indices-and-Methodologies-Lit-Review.pdf, accessed 15 July 2015).

Chellaney, B. (2012). Asia's Worsening Water Crisis. Survival: Global Politics and Strategy, 54(2), 143–156. (Available at http://chellaney.net/2012/03/17/asias-worsening-water-crisis/, accessed 2 July 2015).

FAO (Food and Agriculture Organisation) (2014) Coping with Water Scarcity Challenge of the Twenty-first Century. (Available at http://www.fao.org/nr/water/docs/escarcity.pdf, accessed 2 July 2015).

Rickwood, C.J. and Carr, G.M. (2007). Global drinking Water Quality Index Development and Sensitivity Analysis Report. United Nations Environment Programme Global Environment Monitoring System (GEMS)/Water Programme ISBN 92-95039-14-9.

UNDESA (2014). Water Scarcity. (Available at http://www.un.org/water forlifedecade/scarcity.shtml and references therein, accessed 2 July 2015).

UNWATER (2006). Coping with Water Scarcity. (Available at http://www.unwater.org/downloads/waterscarcity.pdf, accessed 2 July 2015).

Stormwater

Philp, M., McMahon, J., Heyenga, S., Marinoni, O., Jenkins, G., Maheepala, S. and Greenway, M. (2008). Review of Stormwater Harvesting Practices.

Urban Water Security Research Alliance Technical Report No. 9. ISSN 1836-5566 (Online) ISSN 1836-5558 (Print) (Available at http://www.urbanwater alliance.org.au/publications/UWSRA-tr9.pdf, accessed 2 July 2015).

Sustainability

Axis (2014). Thames Water Retrofit Project. (Available at http://www.axiseurope.com/housing_news_1.aspx?news=1:42886&id2=0:37196&id=0:72922&id=0:36052&id=0:36051, accessed 15 July 2015).

Berland, J.M., Estrela, T., Krinner, W., Lallana, C., Leonard, J.W. and Nixon, S. (2001). Sustainable Water Use in Europe – Part 2: Demand Management. (Available at http://edz.bib.uni-mannheim.de/daten/edz-bn/eua/01/Environmental_Issues_No_19.pdf, accessed 2 July 2015).

Burak, S. (2008). Mediterranean Strategy for Sustainable Development Water Use Efficiency: National Study of Turkey. Plan Bleu. (Available at http://www.un.org/esa/sustdev/natlinfo/indicators/egmIndicators/MSSD_latest_eng.pdf, accessed 15 July 2015).

Cole, G., Laffon, L., Krinner, W., Lallana, C., Estrela, T., Nixon, S., Rees, G. and Zabel, T. (1999). Sustainable Water Use in Europe. Part 1: Sectoral Use of Water European Environment Agency.

DEFRA (2011). Water for Life. ISBN 9780101823029.

Jagannathan, N.V., Mohamed, A.S. and Kremer, A. (2009). Water in the Arab World: Management Perspectives and Innovations. The World Bank, Middle East and North Africa (MNA) Region. (Available at http://siteresources.worldbank.org/INTMENA/Resources/Water_Arab_World_full.pdf, accessed 2 July 2015).

Nixon, S.C., Lack, T.J., Hunt, D.T.E., Lallana, C. and Boschet, A.F. (2000). Sustainable Use of Europe's Water? Environmental Assessment Series No. 7. (Available at http://www.eea.europa.eu/publications/water_assmnt07, accessed 2 July 2015).

Phipps, D., Alkhaddar, R., Morgan, R., McClelland, R. and Doherty, R. (2009). Sustainability in Energy and Buildings. Part 6, 357–368, doi:10.1007/978-3-642-03454-1_37.

Singh, K. (2010). Water for Sustainable Urban Human Settlements. UN World Water Assessment Programme. (Available at http://unesdoc.unesco.org/images/0021/002112/211294E.pdf, accessed 2 July 2015).

Water efficiency and saving

Bathroom Manufacturers Association (2010). The Water Efficient Product Labelling Scheme. (Available at http://www.europeanwaterlabel.eu/, accessed 15 July 2015).

Correlje, A.F., de Graaf, R.E., Ryu, M., Schuetze, T., Tjallingii, P. and Van de Ven, F.H.M. (2008). Every Drop Counts – Environmentally Sound Technologies for Urban and Domestic Water Use Efficiency. United Nations Environment Programme, ISBN 978-92-807-2861-3. (Available at http://www.unep.org/ietc/Portals/136/Publications/Water&Sanitation/EveryDropCounts_Sourcebook_final_web.pdf, accessed 2 July 2015).

Critchley, R. and Phipps, D.A. (2007). Water and Energy Efficient Showers, United Utilities. (Available at http://www.allianceforwaterefficiency.org/assets/0/28/142/48/88/C86DEB33-2463-4795-BE5E-A66EA64CAB3E. pdf, accessed 2 July 2015).

EEA (2012). Towards Efficient Use of Water Resources in Europe EEA Report No 1/2012, Copenhagen. ISBN 978-92-9213-275-0. (Available at http://www.eea. europa.eu/publications/towards-efficient-use-of-water, accessed 2 July 2015).

Energy Saving Trust. (2012). Saving Water. (Available at http://www.energy savingtrust.org.uk/domestic/saving-water, accessed 15 July 2015).

Griggs, J. and Burns, J. (2009). Water Efficiency in New Homes – An Introductory Guide for Housebuilders. IHS BRE Press ISBN 978-1-84806-099-9.

Institute for Sustainability. (2014). Delivering and Funding Housing Retrofit: A Review of Community Models. (Available at http://www.instituteforsustain ability.co.uk/uploads/File/Delivering%20and%20Funding%20Housing%20 Retrofit%20report_March%202013.pdf, accessed 2 July 2015).

London Assembly Health and Environment Committee (2012). Water Matters: Efficient Water Management in London. (Available at http://www.london. gov.uk/sites/default/files/Water%20management%20report%20pdf.pdf, accessed 2 July 2015).

London Climate Change Partnership. (2009). Economic Incentive Schemes for Retrofitting London's Existing Homes for Climate Change Impacts. (Available at http://climatelondon.org.uk/wp-content/uploads/2012/10/ Economic-incentive-schemes-for-retrofitting-Londons-existing-homes-for-climate-change-impacts.pdf, accessed 2 July 2015).

McManan-Smith, T., ed. (2007). Water efficient solutions: The Practical Guide for Industry, Commerce and the Public Sector. DEFRA. (Available at http:// wtl.defra.gov.uk/core_files/Water%20Efficient%20Solutions%202007.pdf, accessed 2 July 2015).

Nadel, B. and Butcher, K. (2008). United Utilities Water Efficient Showerhead Offer: Project Report. (Available at http://www.alchallis.com/documents/ UnitedUtilitiesChallisWaterSavingShowerreport.pdf, accessed 2 July 2015).

Schlunke, A., James, L. and Fane, S. (2008). Analysis of Australian Opportunities for More Water-Efficient Toilets. Institute for Sustainable Futures. (Available at http://www.waterrating.gov.au/resource/analysis-australian-opportunities-more-water-efficient-toilets, accessed 15 July 2015).

Sim, P., McDonald, A., Parsons, J. and Rees, P. (2007). WaND Briefing Note 28 Revised Options for UK Domestic Water Reduction. A Review Working Paper 07/04, WaND, University of Leeds. (Available at http://eprints.whiterose. ac.uk/4978/1/Water_Conservation_Lit_Review2.pdf, accessed 2 July 2015).

The Chartered Institute of Plumbing & Heating Engineering. (2014). Retro-fitting for Water Efficiency. (Available at http://www.bre.co.uk/filelibrary/ events/Water%20efficiency/John%27s_presentation_for__wat_F.ppt_ %5BCompatibility_Mode%5D.pdf, accessed 2 July 2015).

UK Green Building Council (2013). Retrofit Incentives. (Available at http:// www.ukgbc.org/resources/publication/uk-gbc-task-group-report-retrofit-incentives, accessed 15 July 2015).

Waterwise (2009a). Evaluation of the Water Saving Potential of Social Housing Stock in the Greater London Area. (Available at http://www.waterwise.org.uk/data/2009_Waterwise_social_housing_stock.pdf, accessed 15 July 2015).

Waterwise (2009b). Water Efficiency Retrofitting: A Best Practice Guide. (Available at http://www.waterwise.org.uk/data/resources/30/water-efficiency-retrofitting-best-practice_final.pdf, accessed 2 July 2015).

Waterwise (2009c). The Water and Energy Implications of Bathing and Showering Behaviours and Technologies. (Available at http://www.waterwise.org.uk/data/resources/27/final-water-and-energy-implications-of-personal-bathing.pdf, accessed 2 July 2015).

Waterwise (2014a). Training and Qualifications (Waterwise CABI Water Efficiency Qualifications). (Available at http://www.waterwise.org.uk/pages/training-and-workshops.html, accessed 2 July 2015).

Waterwise (2014b). Planning Effective Water Efficiency Initiatives. (Available at http://www.waterwise.org.uk/data/resources/51/Planning-effective-water-efficiency-initiatives_Final.pdf, accessed 2 July 2015).

Waterwise East (2010). Water Efficiency in New Developments: A Best Practice Guide. (Available at http://www.waterwise.org.uk/resources.php/28/water-efficiency-in-new-developments-a-best-practice-guide, accessed 2 July 2015).

Water poverty

Consumer Council for Water (2014). Living with Water Poverty: Keeping Your Head Above Water. (Available at http://www.ccwater.org.uk/wp-content/uploads/2014/09/Living-with-water-poverty-in-2014-Report-of-research-findings.pdf, accessed 15 July 2015).

Molle, F. and Mollinga, P. (2003). Water Poverty Indicators: Conceptual Problems and Policy Issues. Water Policy, 5, 529–544. (Available at http://ird.academia.edu/FrancoisMolle/Papers/909082/Water_poverty_indicators_conceptual_problems_and_policy_issues, accessed 2 July 2015).

8 Environmental Law for the Built Environment and Environmental Impact Assessment

Jack Rostron

There is need for international collaboration (i.e. policies, reports, protocols, conferences, laws, regulations and directives) and local actions so that sustainability of the world habitat can be achieved. According to the United Nations scientific panel's report on climate change, collective and significant global action is needed to reduce greenhouse gas emissions to keep global warming below 2°C (EC website, 2014). In accordance with this report, the Intergovernmental Panel on Climate Change's report calls for immediate action as the longer we wait the more expensive and technologically challenging meeting this goal will be (EC website, 2014). The main international collaboration attempts include the Kyoto protocol and the 2012 UN RIO+20 United Nations Conference on Sustainable Development (Tukker, 2013: 272; Dittmar, 2014; UNEP). The success of Rio+20 conferences lies in launching the Global Research Forum on Sustainable Production and Consumption (Vergragt et al., 2014: 1). The international meetings and programmes, such as Rio+20, the Resource Efficiency and Cleaner Production Program, the Ten Year Framework of Programs on Sustainable Consumption and Production, the UN Resource Panel and the Green Economy Initiative programmes, underlined the need for sustainability, sustainable production and consumption as well as for sustainable development. Governments and corporations are being increasingly held responsible for their sustainability performance (Roy and Goll, 2014: 849). The EU policies encourage the reduction in CO_2 emissions especially through the following policies and targets:

- The 25-year Strategic Research Agenda for the European Construction Sector, which aims to achieve sustainability.
- GHG Protocol (2004) by the World Business Council for Sustainable Development and the World Resources Institute and the Carbon Disclosure Project.
- The EC's 20-20-20 targets, an integrated approach to climate and energy policy to fight against climate change.
- The EU framework on climate and energy for 2030 aiming at cost effective reduction in carbon emissions.
- The European Commission's Roadmap for moving to a low-carbon economy in 2050, aiming for the EU to cut its emissions to 80% below 1990 levels through domestic reductions by 2050.

To appreciate the importance of international environmental law, it is necessary to understand the difference between national and international law. National laws are those adopted by national governments of individual countries.

Although such national laws are adopted by individual countries they may have international implications. A foreign manufacturer or producer may cause loss to a person in another country. The manufacturer or producer may still be held liable for the losses suffered in the affected country. While such laws affect international activities they are generally not considered international law.

International law, on the other hand, concerns agreements or treaties between different countries. The growth of international law in the environmental area can be considered to have been begun in the 1970s with the Stockholm Conference on the Environment in 1972. Since then interest has grown considerably.

Areas of concern in which international laws have or are being evolved include: ozone layer depletion and global warming, desertification, destruction of tropical rainforests, marine plastics pollution from ships, international transportation of hazardous substances and so on.

International co-operation in the form of agreements, treaties and resolutions created by intergovernmental organisations as well as national laws and regulations are being developed by such organisations as: United Nations Environment Programme (UNEP), the European Union, the OECD and the Council of Europe.

There are vehicles for implementing and enforcing international environmental disputes. These forums, however, generally require the parties to voluntary agree to place the dispute before an appropriate court, tribunal or panel. Even when these forums obtain the agreement of the parties to adjudicate international environmental disputes, they still need in most instances to rely on the co-operation of national governments to enforce judgments. Such co-operation, however, is often not forthcoming for political or commercial self-interest reasons.

There are a small number of environmental agreements that have led to the establishment of international institutions that can directly impose trade sanctions, such as the Montreal Protocol. The types of sanctions envisioned under the Montreal Protocol are procedurally difficult to impose. In general there is no international body authorised to directly enforce international environmental law.

Countries often try to accept or avoid international environmental obligations because it is in their economic self-interest to do so. It would be unusual for national governments to enter into agreements which may harm their own countries' economic well-being for altruistic reasons. Such actions as these are taken under expectations of political or economic benefit sooner or later.

However, ultimate responsibility at the moment for the protection of the environment remains with national governments and individual jurisdictions. It can be considered that there is a need for the establishment of an international court of environmental jurisdiction. At the moment, there would appear little movement towards establishing such a vehicle, even though the need for sustainable development is recognised by most countries around the world.

For sustainability to be successful the law surrounding the planning process must foster development that is sustainable and discourage that which is unsustainable. Perhaps the main vehicle for achieving this aim is the concept of environmental impact analysis. This chapter reviews the development of planning law concerning environmental impact assessment in the UK, as a case study, describing the agencies involved and the regulatory framework underlying such assessments. The procedures required are discussed along with the underlying concept of sustainable development. The technical process involved in undertaking an environmental impact assessment is described in detail.

8.1 Sustainable development and EIA decision making

The concepts of sustainability and development of environmental impact assessment techniques are closely interrelated. The two topics are not only intertwined as concepts, but the geographical distribution of environmental pollutants requires transnational administrative protocols and wide legal jurisdictions. In recent times this has resulted in the creation of bodies such as the European Environment Agency. The evaluation of environmental impact, in terms of the breadth and depth of topics covered, is clearly related to the perceived level of environmental impact damage. There is clearly a need for the sustainability of development to be considered in the formal context of statutory environmental impact.

There is no universally accepted definition of sustainable development (Pearce, 1993). The definition most commonly used is that of the Brundtland Commission, which stated that sustainable development "meets the needs of the present without compromising the ability of future generations to meet their own needs" (World Commission on Environment and Development, 1987). Sustainable development generally encompasses three core ideas:

- the need to embed environmental considerations into economic policy making,
- a commitment to equity (i.e. fair distribution of resources, including intergenerational equity),
- a notion of economic welfare – that is, development (not growth) that acknowledges non-financial components, such as health, education and the environment.

For proposed development to be sustainable, it clearly needs to be acceptable in terms of pervasive environmental standards. Towards this end, the tool of environmental impact assessment assists decision makers in the planning process in reviewing the potential sustainability of particular development proposals.

Planning is not just about land use; there are also issues of policy involved. Policy is being driven at present by the idea of sustainability, which grew out of earlier debates on limiting growth to avoid outrunning Earth's ecological resources (Meadows et al., 1972).

The original "limits to growth" concept was heavily criticised, but it did challenge the prevailing assumption at the time that there were no constraints on continued development. Many environmentalists predicted that continuation of prevailing growth trends would lead to widespread resource scarcity, high levels of pollution and famine on a catastrophic scale. In a more sustainability-conscious world EIAs are, therefore, increasingly important in ensuring that development proposals do not infringe environmental standards adopted in local policy planning documents.

Current patterns of economic growth do cause major ecological problems, but this might change with the use of new technology, which could stimulate growth while minimising environmental impacts. Even growth in consumption of physical resources need not be reduced to zero to be sustainable. The problem is not necessarily growth itself, but rather the rate of growth (Jacobs, 1991).

Environmental impact assessment is a systematic procedure for considering the possible environmental effects of any proposed development before a decision is made about whether the project should be given approval to proceed. More fully, EIA refers to the entire process and all the techniques by which the impact of a proposed development on the environment can be assessed.

8.1.1 Environmental impacts

Environmental impacts are effects caused by the development project, especially negative effects that may lead to deterioration in or destruction of the existing environment. Impacts can be direct (primary) or indirect (secondary). They can be of short, medium or long-term duration and may be temporary or permanent. Environmental impacts can include positive effects, although the EIA process mainly focuses on negative impacts. Impacts may be cumulative. Often, this cumulative effect is simply additive. However, it can be synergistic in cases where, for example, several different pollutants create multiplier effects. In some cases, one pollutant might interact with another, causing an antagonistic effect. This situation requires particular attention and specific assessment in order to understand the effects involved and ensure that the expected mitigation result is achieved. Major projects can give rise to a wide range of impacts. Because environmental systems do not exist in isolation, it is essential that all impact possibilities be assessed. For each effect or likely impact, there are certain areas that need to be reviewed in the context of the EIA.

8.1.2 Environmental impact assessment

Directive 85/337/EC (dealing with assessment of the environmental effects of certain public and private projects) created the basic requirements for environmental impact assessment (EIA). Alder (1993), Boch (1997), Carnwath (1991) and Jones et al. (1998) provide detailed information on the environmental impact assessment. In cases where EIA is required, planning permission cannot be granted unless environmental impact assessment has been carried out and the results have been taken into account. Schedule 1 of the applicable regulations lists the types of projects for which EIA is always required (see Directive 97/11/EC, amending Directive 85/337/EEC). These are larger projects that involve any of the following elements: crude oil refinery; thermal or nuclear power station; radioactive waste storage facility; integrated chemical installation; special road; trading port; waste disposal installation for the incineration or chemical treatment of special waste (for a full list, see Schedule 1 to the regulations). Schedule 2 contains a list of projects that require EIA if they will have a significant effect on the environment due to their size, location or nature (as determined on a case-by-case basis by the LPA or the developer). They include projects in the following industries or with certain specified characteristics: agriculture (e.g. poultry or pig rearing, salmon farming); extractive industries (e.g. extraction of peat, minerals, coal or natural gas); energy (e.g. power stations); processing of metals (e.g. ship yards, iron works or car plants); chemicals; food; textiles, leather, wood and paper; rubber; infrastructure (e.g. industrial estate development, roads, harbours, dams, tramways, yacht marinas and motorway service areas) and certain other projects (e.g. holiday villages, hotel complexes, knackers' yards, wastewater treatment plants, sites for depositing sludge or installations for the disposal of controlled waste or water from mines and quarries not specified in Schedule 1).

The following checklist details the issues that should be considered for each significant impact:

- *Scientific Background*: It is necessary to understand the basic science and technology related to each impact. For instance, in the case of noise impacts, it is essential to know how the noise will be generated, how it can be attenuated and the effects of noise on the environment.
- *Legislative Background*: Many aspects of EIA involve their own specialist areas of legislation. The field of environmental law is burgeoning, especially given continual UK legislation in this area and ongoing interest by the European Union. Consider, for example, the field of water quality. There are a multitude of statutory enactments and standards governing both fresh water and wastewater. An EIA would need to take all these requirements into consideration.
- *Impacts and Interest Groups*: For each aspect of the environment studied, there are likely to be a number of interest groups who will have a view on the impacts of the development. These groups typically vary in size, ranging from large national or international organisations down to very local groups. Their range of interests may also vary from general to specialised.
- *Baseline Studies*: Before any action can be taken on an EIA, it is essential that the state of the existing environment be understood. Thus, appraisal of each potential environmental impact should begin with a baseline study. It is essential for the developer to carry out an audit or inventory that completely describes the current state of the environment into which the project will be placed. Fortunately, there is an enormous amount of information available about the environment in the UK and Europe. It can be obtained from a range of sources, including governmental and quasi-governmental organisations, LPAs, water companies, heritage organisations, interest groups and local organisations.
- *Impact Prediction*: Once the existing environment is understood, the next step is to predict the likely impacts of the proposed development, in both qualitative and quantitative terms. The developer, as the party who is most familiar with the likely effects of the project, must assess the significance of these impacts.
- *Dealing with Impacts That Are Identified*: Once the significant impacts are identified and quantified for each aspect of the environment, it is then necessary to consider how to mitigate these impacts. In order to ensure the success of the mitigation measures adopted and the overall management of the project's environmental impacts, it is also essential to include provisions for monitoring. Many aspects of the EIA process are still relatively new and innovative. If the process is to work well and move forward, a commitment to monitoring by both developers and LPAs is fundamental.

- *Mitigation*: Mitigation refers to measures the developer proposes to take in order to prevent, avoid or ameliorate the actual or potential environmental impacts of the proposed project. Mitigation measures relate to the significant impacts detailed in the environmental statement. They might include substitution of techniques, use of cleaner methods, pollution control measures, compensation or attenuation of pollutants. Decisions on mitigation measures are an integral part of the design and planning of any project. This is an interactive process. As the project design develops and information becomes available, feedback loops will be needed to modify various facets of the design, including mitigation. Mitigation measures can be of several types: avoidance of impacts, reduction of impacts or remedies that help counteract impacts. When deciding on mitigation methods, two principles are central to European Commission policy on the environment:

  ○ Preventive action is better than remedial measures.
  ○ Where possible, environmental damage should be rectified at the source.

Given this background, the best form of mitigation is modification to the project to avoid impacts, rather than containment or repair. Some mitigation measures are required by law in the case of certain new facilities (e.g. flue gas desulfurisation systems for power plants). One key question for the developer is how far the mitigation measures should go. Should mitigation be total, with no regard to cost? To some extent, this issue has been settled by "best practicable environmental option" permitting, which aims to limit damage to the environment at a cost that is deemed "acceptable". For the developer, the ultimate goal is to provide enough mitigation measures to ensure that its planning application will succeed. The overall objective of EIA is not to prevent development schemes from going ahead, but to mitigate them to a point where they are environmentally acceptable. For the developer, however, the question of cost will always be significant. Expensive mitigation measures might make the project financially impractical.

8.2 Case study: UK

The issue of sustainability currently is being developed within several different contexts – international (Boyle and Freestone, 1999), European and UK-wide. The importance the UK government places on developing sustainable policies is clear from the wide range of initiatives being pursued in this area and the number of entities that have been set up to review these issues.

The UK government has laid down four broad objectives for sustainable development: social progress, which recognises the needs of all members of society; effective protection of the environment; prudent use of natural resources; as well as maintenance of high and stable levels of economic growth and employment, so that everyone in Britain can share in high living standards and greater job opportunities (Department of the Environment, Transport and the Regions, 1999).

The European Environment Agency publishes regular indicator reports on the state of the European environment (European Environment Agency, 1998). Its report on the UK, entitled "Setting Environmental Standards", was published in October 1998. (See also Royal Commission on Environmental Pollution (1997). *21st Report, Cmnd 4053*. London: HMSO.) Such reports have indicated the increased need for EIAs to be made in the statutory planning process. For this reason, these reports highlighted the importance of EIAs for sustainable development. In accordance with the European Environment Agency's report, the UK also prepares its regular reports on sustainability, sustainable development and environment. The UK Round Table on Sustainable Development has recommended that sustainable development reporting should: include a number of different reports, rather than just one annual report; be targeted at different audiences; use a number of different formats, including information technology; be open and transparent in order to create trust among members of the public and be an interactive process that includes not only report publication, but also feedback and re-examination of both sustainability targets and the reporting process itself (Department of the Environment, Transport and the Regions, 1998).

In *Quality of Life Counts* (Department of the Environment, Transport and the Regions, 1998), the UK government listed 15 headline indicators that are intended to give a broad overview of sustainability trends and 150 national indicators that look at specific issues and identify areas for action. In relation to the environment, the headline indicators include: climate change, air quality, road traffic, river-water quality, wildlife and farmland, land use, household waste and other waste.

8.2.1 Planning controls and environmental protection in the UK

Town planning was first introduced in the Housing Town Planning Act 1909. It was followed by the Housing Act 1923, which provided that account should be taken of the architectural and historic characteristics within an area. This provision was strengthened in the Town and Country Planning Acts 1932 and 1944. However, it was the Town and Country Planning Act 1947 that genuinely put planning and conservation into place and set the scene for the modern system of planning we have today (Hughes, 1996). This law was modified and consolidated in the Town and Country Planning Act 1990. The following related legislation reinforced the step change in planning law: the Planning and Compensation Act 1991, the Planning (Listed Buildings and Conservation Areas) Act 1990 and the Planning (Hazardous Substances) Act 1990, Planning (Hazardous Substances) Regulations, SI 1992/596, SI 1995/226 and Town and Country Planning, England and Wales: The Planning (Control of Major Accident Hazards) Regulations 1999. There are also a large number of departmental circulars that provide guidance on applying the applicable statutes.

Planning permission is required for "developments" and granted by the Local Planning Authority (LPA). Development is defined in Section 55 of the Town and Country Planning Act 1990 (TCPA) as the carrying out of the following operations:

- Engineering: Operations "usually carried out by engineers".
- Mining.
- Building: Under Section 55(2), this includes demolition, rebuilding, structural operations and any operations normally undertaken by a builder. It does not include interior work.
- Material change of use: A material change to land or buildings.

In reaching a decision on a planning permission request, the LPA must review the area's local development plan and make a determination in accordance with it unless material considerations indicate otherwise (TCPA 1990, Section 54A). A development plan sets out the specific planning policies and proposals of local authorities for each particular area. The local plan forms the basis for future planning applications. Local development plans help ensure that environmental protection is considered at the level of policy making (Purdue, 1994; Herbert-Young, 1995; Wood, 1996). Although these plans set down the basic ground rules for development within a locality, they increasingly are being overtaken by central government circulars and policy guidance notes. Development plans often allocate land to particular residential and commercial uses such as retail, industrial and so on. They also can include environmental measures, such as provisions to limit traffic-related pollution by means of a commitment to take particular actions, such as developing green corridors.

The use of development plans to ensure environmental protection has been given more prominence through regulations and guidance. The National Planning Policy Framework 2012 replaced much of the detailed guidance contained in the government's former guidance and policies to LPAs. The framework sets out the government's current requirements for the planning system in England. It sets the framework for how LPAs should produce their development plans. The following rules generally govern LPA planning permission decisions:

- LPAs can grant permission with conditions such as time limits for completion of building works. These generally can encompass whatever provisions the LPA considers fit. Any condition imposed must be for a planning purpose, must fairly and reasonably relate to the development and must not be unreasonable (see *Newbury District Council v Secretary of State*, (1981) AC 578).
- Under section 106 of TCPA 1990, the LPA may enter into an agreement providing for "planning gain" (Arnold, 1999) that relates to carrying out certain activities or refraining from using the land. Planning gain often involves the applicant in paying funds to the local planning authority for the provision of facilities such as affordable housing or highway works. Planning gain is similar to "impact fees" in the United States, but tends to be broader in scope and can be used for a wider range of purposes. The planning gain must be capable of being a material consideration; it also must be necessary and relevant to the proposed development (e.g. Department of the Environment Circulars 16/91 and 11/97 (Planning Obligations). In the case of the *Tesco Stores Ltd v Secretary of State for the Environment*, (1995) vol. 2 All ER 636, the court stated that the test is whether the obligation has some connection with the development; situations where planning gain is offered and where it is required, should be distinguished).
- Planning permission can be granted unconditionally or refused.
- An applicant can appeal within six months if permission is refused. The applicant can also object if a condition is imposed (e.g. according to *Newbury District Council v Secretary of State*, (1981) AC 578, a condition imposed for a planning purpose must relate to the development and must not be manifestly unreasonable) or if the local authority does not make a decision within eight weeks. Such an appeal must be made within 28 days.

- The appeal is heard by a planning inspector, who may allow, dismiss or vary the LPA decision. The appeal may be made by means of an oral hearing in front of the planning inspector or by written representations.
- The Secretary of State for Communities and Local Government can "call in" (remove from an LPA's jurisdiction) a planning application under TCPA 1990, Section 77, in cases that involve issues of more than local importance, regional or national controversy or conflict with national policy. The Secretary of State will "call in" an application because the proposal is of national importance or too complicated to be dealt with by a local municipality and consequently must be considered by central government.
- A judicial review application may be brought against the LPA or the Secretary of State if an applicant believes that a decision has been made unfairly. Recourse to the local or central government ombudsman is also available in cases of maladministration (Scrase, 1999).

In a planning inquiry, an inspector is appointed to hear evidence from those who support and object to a proposed development project. Often an inquiry leads only to a recommendation that may or may not be accepted. A planning inquiry is a mechanism for hearing public views – and allowing the government to be seen as listening to those views. If the inspector recommends that the proposed project go ahead, this decision provides the government with support in the form of an independent inquiry. If the inspector recommends abandoning the project, the government may face political problems if it allows the project to proceed. Planning inquiries can vary in scope and function. Some may concern large projects, which affect whole communities or large geographical areas. Others may affect only one street or housing estate. Planning inquiries aim to provide government with all the facts so that a properly informed decision can be made. Traditionally, inquiries have been regarded as "merely a stage in the process of arriving at an administrative decision", for example *Johnson and Co. (Builders) Ltd. v Minister of Health*, (1947) 2 All ER 395. Inquiries into objections are usually mandatory and tend to cover many areas of the planning process. Most inquiries examine local objections to a particular proposal or decision. However, a major inquiry may review issues that involve not just local interest, but national or regional importance. In many cases, the power to convene an inquiry is conferred by statute on the Secretary of State, for example, Town and Country Planning (Inquiries Procedure) Rules 1992/2038. These inquiries are bound by statutory rules (derived from Section 78 of TCPA 1990) that set out their basic procedure. Such inquiries fall into two categories, each with its own set of rules:

- recovered cases (where the Secretary of State makes a decision after the recommendation of the inspector);
- transferred cases (where the inspector makes a decision on behalf of the Secretary of State) Town and Country Planning (Determination of Appeals by Appointed Persons) (Prescribed Classes) Regulations 1999/420 extended the transferred cases procedure to a wider number of cases.

In the case of a major inquiry, there is a special process, which the Secretary of State can follow if it is found to be desirable. It allows the inspector, under certain conditions, to make a decision without all the evidence being available. Only the applicant can appeal against a planning decision. But once the applicant has done so, other parties who have made representations within the specified time limits may have a right to be heard at the inquiry. This right may be subject to the discretion of the inspector.

8.2.2 Environmental impact assessment in the UK[1]

In cases where EIA is required, planning permission cannot be granted unless environmental impact assessment has been carried out and the results have been taken into account. Projects that require EIA have been listed in Schedules 1 and 2 of the Directive 97/11/EC, amending Directive 85/337/EEC as mentioned previously in this chapter. The Town and Country Planning (Environmental Impact Assessment) (England and Wales) Regulations 1999, SI 1999 No. 293, replaced earlier regulations dating from 1988 (Circular 2/99 gives guidance on the 1999 regulations). The amended regulations enlarged the range of projects that are subject to environmental impact assessment and made some procedural changes, including the following:

- Regulation 10 allows developers to obtain a formal initial opinion from the relevant planning authority regarding what should be included in the environmental statement.
- For all Schedule 2 developments (including those which would otherwise benefit from permitted development rights), the LPA must adopt its own formal determination (or "screening opinion") as to whether EIA is required. This can be done before (regulation 5) or after (regulation 7) a planning application has been submitted and placed on the planning register.
- Part II, paragraph 2 of Schedule 4 now requires a developer to include in the environmental statement an outline of the main alternatives considered and the reasons for the choices that were made.
- Under regulation 21, when making a determination on an EIA application, the LPA or Secretary of State must inform the public of its decision and the main reasons for it, including whether the application was granted or refused.

A developer can apply to the LPA for an opinion as to whether environmental impact assessment is needed for a particular project (e.g. *R v Rochdale Metropolitan Borough Council ex parte Tew*, (1999, July) Environmental Law Bulletin at 12 [in relation to the EIA and outline applications]). The LPA has to respond within three weeks after it receives the request. If the LPA decides that environmental impact assessment is required, but the developer disagrees, the developer can appeal to the Secretary of State. The Secretary will generally try to issue a direction on the case within three weeks. There is no appeal from the Secretary of State's decision. Third parties who believe that a proposed project requires an EIA may contact the relevant LPA to set out their views. In addition, the Secretary of State can issue a direction at any time regarding whether an EIA is needed for a particular project.

The following paragraphs offer a brief summary of environmental impact assessment procedure in the UK.

- *Developer's Initial Steps*: At the start of any project affected by the EIA process, the developer should carry out detailed surveys of the site where the project is to be located. It should also seek to "design out" any adverse environmental impacts.

- *Submission of the Environmental Statement*: As part of the EIA procedure, the developer must submit an environmental statement to the "competent authority". This will be the LPA if planning permission is also required. In other cases, it will be the public body with responsibility for the particular area at issue (e.g. the Forestry Commission). The environmental statement should record the developer's assessment of the project's likely effects on the environment and how those effects will be mitigated. The statement should identify all potential environmental effects and specify steps that are envisaged to avoid, reduce or remedy any adverse effects. The environmental statement must contain the following information: a description of the proposed development and site design; data needed to evaluate the environmental effects (both direct and indirect) of the project; a description of measures proposed to mitigate significant adverse environmental effects from the project and a summary written in non-technical language.

- *Consultation and Public Input*: Once the environmental statement is submitted, the relevant authority consults with other interested entities (such as English Nature and other groups with specialist knowledge) on the content of the statement. Members of the public can also comment on the statement. The development proposal must be publicly advertised with details about where copies of the environmental statement can be obtained or reviewed. A minimum notification period of 21 days typically is required.

- *Outcome of Environmental Impact Assessment*: The relevant authority must complete environmental impact assessment on proposed developments before deciding whether they may go ahead. The authority should be fully informed about all environmental implications of the development and must take these into account when making its decision. The assessment should also consider the views of the developer, the public and consultees. Even if the EIA finds that a particular development will have adverse environmental effects, this does not automatically mean that the project will not be allowed to go forward. The authority must also consider other relevant factors, such as economic, social, public health and safety issues and any suitable mitigation measures that may be imposed.

As English legal law system is common law and case based, disputes in EIA practice will establish precedents for resolving future disputes. Key court decisions addressing EIA issues in the UK include the following cases:

- The adequacy of the EIA procedure was reviewed in *R v Poole Borough Council, ex parte Beebee and Others* (JPL 643, 1991). In this case, Poole Borough Council gave itself planning permission to construct houses on Canford Heath, a Bronze Age heath land that contained several rare wildlife species, including newts, toads, the Dartford Warbler and the Dorset Blue Butterfly. The court held that the LPA did not have to undertake environmental impact assessment in this case because the authority already had all the relevant information before it. This decision was later reversed by Michael Heseltine, who was then serving as Secretary of State for the Environment, Transport and the Regions.

- Similarly, in *Twyford Parish Council v Secretary of State for the Environment, Transport and the Regions* (4 JEL 273, 1992), the court refused to intervene in a decision regarding development of a controversial section of roadway even though the decision had been made illegally, noting that the road had already been delayed for long enough and the applicants had not suffered from harm.

- In *R v Swale Borough Council ex parte RSPB* (1 PLR 6, 1991), it was held that local planning authorities have broad discretion to decide whether projects need environmental impact assessment and courts should interfere with that discretion only in exceptional cases (see *R v North Yorkshire County Council ex parte Brown*, (1999) 2 WLR 452 [court held that EIA was needed when an old mining consent was renewed with new conditions placed upon it]).

The following discussion aims to explain the process of EIA within the framework of the applicable UK legislation and regulations. It first offers a definition of environmental impact assessment and then discusses the stages involved in the EIA process in more detail. Any environmental impact assessment will involve many complex aspects. Within each of these aspects are numerous elements relating to science, law and public policy. Explaining all these elements is beyond the scope of this discussion. Instead, this coverage endeavours to give an overview that describes the EIA process and the complications that participants in the process might have to face.

The EIA process in the UK consists of several stages. The process is a series of iterative steps, with each step potentially feeding back into previous stages. Initially, the developer collects information from available sources about the environmental effects of its proposed project. This information is presented to the LPA (the Local Planning Authority), which takes it into account in deciding whether the proposed project should be allowed to proceed. EIA is intended to be a systematic and comprehensive assessment of significant impacts. It involves participation by the project developer, the LPA, other sources who might have relevant information and the public. The developer presents information on the proposed project to the LPA in an environmental statement and to the public in a non-technical summary. Based on the environmental statement, the LPA evaluates the project's significant effects (and the scope that may be available for modifying or mitigating them) in an environmental appraisal before making a decision.

Consideration of Alternatives: In this step, the LPA considers alternatives related to the proposed project. The alternatives considered involve a range of factors, including project location, processes and configuration. The "no action" alternative (i.e. the option of not proceeding with the project) must also be considered. Current regulations require this step to be documented.

Project Design: EIA should be carried out in parallel with the project design process. This allows the EIA process to contribute input regarding ways the project's design can be modified (or even changed completely) to create the "best practicable environmental option". Pursuing the EIA process in conjunction with project design generally is more efficient for the developer, which can save money by incorporating changes before the design is completed, while also avoiding lengthy delays in the planning process.

Screening for Significant Environmental Impacts: This stage in the process is intended to determine whether a full EIA is necessary for the project concerned. Screening allows the authority to focus environmental impact assessment on those projects that are likely to have significant environmental effects. Whether a particular project requires EIA is determined by the applicable screening criteria and thresholds. In addition, it depends in part on how (and whether) the LPA applies the appropriate criteria. The applicable UK regulations contain two screening schedules. Schedule 1 applies to projects for which EIA is required in every case. Schedule 2 applies to projects for which EIA is required only if the particular project is judged likely to give rise to significant environmental effects. Determinations about the significance of environmental effects are likely to hinge on the nature, size and location of the particular project:

- projects that are likely to give rise to complex or particularly adverse impacts (e.g. polluting discharges) are more likely to be considered significant in their impacts
- larger sized projects are also more likely to have significant impact, particularly if the project is of more than local importance
- projects intended for sensitive locations always require EIA; sensitive locations include sites of special scientific interest, areas of outstanding natural beauty, local nature reserves, national parks and similar locations.

The screening process has been formalised by the applicable regulations. The developer can obtain a screening opinion from the LPA or a screening directive from the Secretary of State. In practice, nearly all types of public and private projects are subject to EIA. The original regulations excluded certain projects (such as private motorways, wind generators and coastal protection projects). However, these were later included by the Planning and Compensation Act 1991, which specifies that the Secretary of State is allowed to require EIA for them.

Scoping EIA Coverage: Scoping is the process of deciding on the aspects to be considered in the EIA. This step narrows the assessment down to the likely most significant impacts and defines those areas of the environment to be studied. Each project (and the environment into which it will be placed) is different. Thus, a necessary antecedent to each EIA process involves deciding what to cover in the exercise. Scoping at an early stage can be more efficient for the developer. It can also help focus the environmental statement and make it much more effective in the decision-making process.

The Environmental Statement: This is a statement setting out environmental information about the proposed project. The developer must submit it with the planning application.

- *Assessment Team*: The developer is responsible for submission of the environmental statement and for successful execution of the EIA process. The task typically is complex, requiring many different types of specialist skills and knowledge and the timescale is limited. As a result, the assessment usually is handled by a developer-led team. For a smaller EIA (such as a straightforward single-purpose Schedule 2 project), most of the work would probably be done by one firm of consultants, together with the appropriate specialists as required by the nature of the project. For larger, higher profile projects (such as the Channel Tunnel, the Channel Tunnel rail link or a new airport development), a consortium of consultant firms might be required. These consultants would in turn consult with the relevant specialist professions. An example of the scale of this consultation can be gauged from the fact that one of the consultants engaged for the Channel Tunnel EIA consulted with 48 different organisations. Among the types of specialists who could potentially be involved in environmental impact assessment are town planners, civil engineers, statutory authorities, architects, landscape architects, agricultural scientists, economists, geologists, hydrologists, archaeologists, sociologists, chemists, ecologists and meteorologists.

- *Crafting the Environmental Statement*: The environmental statement sets out the developer's own assessment of the proposed project's likely significant environmental impacts. It also includes other relevant information, such as mitigation measures. The environmental statement is prepared by the developer, who is required by regulation to submit it, along with his application, to the LPA in order to obtain planning consent. The environmental statement can be thought of as the report of the EIA process. The aim of the report is to give interested parties an accessible document that details the assessment and gives due weight to all significant environmental effects or impacts. The LPA may express opinions about the content of the environmental statement and it has the power to call for further information if necessary. Any disagreements about the environmental statement can be addressed through the planning appeals procedure. The developer will usually engage consultants for all or part of the EIA process, including preparation of sections of the environmental statement. The preparation of the statement, if it is to be effective, should involve collaboration among the developer, the LPA and the various consultees. There is no prescribed form for the environmental statement, other than the requirement that it satisfy the applicable regulatory mandates. The best format will offer a systematic and objective account of the significant environmental impacts that may result from the project.

- *Consultation and Participation*: When EIA is required, there must be participation by and consultation with a range of stakeholders, including the general public. Some of the entities to be consulted have statutory responsibilities related to aspects of the environment. The LPA is legally required to consult with the specified parties and each must be provided with a copy of the environmental statement for the project. It is preferable for the developer to involve consultees as early as possible after discussions on scoping with the LPA. The developer has to assess the local environment where the project will be located and the consultees have data about this environment, so involving them early on makes for a more efficient process and a more effective environmental statement. Statutory consultees are obliged to provide information for the EIA process. They are required to provide only information already in their possession and are not obligated to undertake original research. They also can charge a fee for this information. Statutory consultees are not obligated to give the developer an opinion on the proposed project, although they have an opportunity to do this at a later stage in the EIA process. In addition to statutory consultees, the developer should consult with the general public and with local organisations during the preparation of the environmental statement. This will give the developer an indication of which issues are likely to be crucial in the assessment process, especially if the application is likely to go to public inquiry. There are many methods of consultation available, some formal and some informal. They include: questionnaires and surveys; advertisements in the media; displays of information; community outreach; workshops; public meetings. All of these methods can be used to elicit response from the public and all might be useful at different times in the process. In addition to being a legal obligation, consultation can be a very useful source of information. If used properly, it can enhance the EIA process.

8.3 Summary

The preceding discussion highlighted the international dimension of environmental impacts and sustainability of development proposals. This is particularly well illustrated by climate change, which has evolved as a global phenomenon with implications for sustainability of developments that transcends national jurisdictions.

If the development is allowed to proceed, it is important to ensure that its environmental impacts are minimised. Mitigation measures should be monitored to ensure their effectiveness.

The overall EIA process should also be audited in order to compare predicted and actual impacts. This is an essential part of the learning curve in environmental management. Audit results should be used to inform future EIA processes.

Environmental impact assessment is not a single specific analytical method or technique. Instead, it involves a combination of methods, techniques and scientific approaches for identifying, predicting and evaluating the environmental impacts likely to be created by particular developments.

The trend in EIA methodology is towards an increasing focus on indirect, cumulative and trans-boundary impacts. Thus, EIA might look closely at the possibility of pollutants being transported from one country to another (e.g. acid rain from the UK being transported to Norway or being deposited into the rivers of Germany).

Considerable research on these topics has taken place in recent years. As a result, much larger databases are now available. In addition, geographic information systems and computer models continue to be developed.

Well-established methods and techniques are now available for analysing the existing environment and predicting the impacts of specific processes. Expert opinion is also readily available in many cases. Among the more widely used tools are traffic surveys, noise surveys and computer models for impact prediction. Databases of knowledge are also available on factors such as noise sources and pollutant attenuation. Using these tools and knowledge sources, a surveyor can value land or buildings and a botanist can quantify the rarity of a plant species.

Many environmental aspects and impacts are also subject to limits specified by statute, regulation or other accepted standards. Among these are, for instance, statutory controls on noise, European Commission directive standards on wastewater quality and discharges and World Health Organization standards on drinking water quality.

Although these standards are based on science, their interpretation is not always totally objective. They can also arouse intense differences of opinion. This is particularly true of visual impacts. It can also be true of social effects and impacts on local environmental resources.

In such cases, no generally recognised standards can be applied to determine the "acceptable" level of impact. Instead, the developer must be prepared to work with the appropriate consultees and explore workable options.

The concepts of sustainability and development of environmental impact assessment techniques are closely interrelated. The two topics are not only intertwined as concepts, but the geographical distribution of environmental pollutants requires transnational administrative protocols and wide legal jurisdictions. In recent times this has resulted in the creation of bodies such as the European Environment Agency. The evaluation of environmental impact, in terms of the breadth and depth of topics covered, is clearly related to the perceived level of environmental impact damage. There is clearly a need for the sustainability of development to be considered in the formal context of statutory environmental impact.

The next chapter draws together the material from all chapters covering managerial, technical and legal aspects of the sustainable built environment.

Acknowledgement

Mr Jack Rostron wishes to thank to Deborah Legge, LLB, MA, PhD, former senior lecturer at the Liverpool John Moores University, who contributed to the research discussed in this chapter.

Notes

1   See Alder (1993); Boch (1997); Carnwath (1991); Jones et al. (1998).

References

Cases

[1991] JPL 643.
[1991] 1 PLR 6.
[1992] 4 JEL 273.
B Johnson and Co. (Builders) Ltd. v Minister of Health, (1947) 2 All ER 395.
Newbury District Council v Secretary of State, (1981) AC 578.
R v Rochdale Metropolitan Borough Council ex parte Tew, (1999, July) Environmental Law Bulletin at 12 (in relation to the EIA and outline applications).
R v North Yorkshire County Council ex parte Brown, (1999) 2 WLR 452.
Tesco Stores Ltd v Secretary of State for the Environment, (1995) vol. 2 All ER 636.

Environmental policies and reports

EC (European Commission) website. (2014). (Available at http://ec.europa.eu/clima/news/articles/news_2014041401_en.htm, accessed 3 July 2015).
EU Policy. (Available at http://ec.europa.eu/clima/policies/package/index_en.htm, accessed 3 July 2015).

Environmental impact assessment

Alder, J. (1993). Environmental impact assessment the inadequacies of English law. Journal of Environmental Law, 5(2), 203–220.

Jones, C.E., Wood, C.M. and Dipper, B. (1998). Environmental Assessment in the UK Planning Process: A Review of Current Practice. Town Planning Review, 69(3), 315–339.

Laws, regulations and directives

Boch, C. (1997). The Enforcement of the Environmental Assessment Directive into the National Courts: A Breach in the Dyke? Journal of Environmental Law, 9(1), 119–138.

Department of the Environment Circulars 16/91 and 11/97 (Planning Obligations).

Department of the Environment, Transport and the Regions (1998, November 2). Press Release 7/98.

Department of the Environment, Transport and the Regions (1998, November 21). Quality of Life Counts: Indicators for a Strategy for Sustainable Development for the UK a Baseline Assessment 1999.

Department of the Environment, Transport and the Regions (1999, May). A Better Quality of Life a Strategy for Sustainable Development (Cmnd. 4345).

Directive 97/11/EC, Amending 85/337/EEC, was implemented in March 1999. See Town and Country Planning, Assessment of Environmental Effects Regulations, SI 1999/293.

Directive 97/11/EC, Amending Directive 85/337/EEC.

Herbert-Young, N. (1995). Reflections on Section 54A and "Plan-Led" Decision-Making. Journal of Planning and Environmental Law, 24, 292–305.

Hughes, D. (1996). Environmental Law, 3rd edition. London: Butterworths.

Planning (Hazardous Substances) Regulations, SI 1992/596. SI 1995/226 and Town and Country Planning, England and Wales: The Planning (Control of Major Accident Hazards) Regulations 1999.

Purdue, M. (1994). The Impact of Section 54A. Journal of Planning and Environmental Law, 23, 399–407.

Royal Commission on Environmental Pollution. (1997). 21st Report, Cmnd 4053. London: HMSO.

TCPA 1990, Section 54A.

Town and Country Planning (Inquiries Procedure) Rules 1992/2038.

Town and Country Planning (Determination of Appeals by Appointed Persons) (Prescribed Classes) Regulations 1999/420 Extended the Transferred Cases Procedure to a Wider Number of Cases.

Planning

Arnold, C. (1999). Planning Gain: How are Off-Site Liabilities Passed Back to Landowners? Journal of Planning and Environmental Law, 29, 869–877.

Carnwath, R. (1991). The Planning Lawyer and the Environment. Journal of Environmental Law, 3(1), 57–67.

Wood, M. (1996). Local Plans and Unitary Development Plans: Is There a Better Way? Journal of Planning and Environmental Law, 25, 807–815.

Sustainability and sustainable development

Barr, S., Gilg, A. and Shaw, G. (2011). Citizens, Consumers and Sustainability: (Re)Framing Environmental Practice in an Age of Climate Change. Global Environmental Change, 21, 1224–1233.

Boyle, A. and Freestone, D., eds. (1999). International Law and Sustainable Development: Past Achievements and Future Challenges. Oxford: Oxford University Press.

Dittmar, M. (2014). Development towards Sustainability: How to Judge Past and Proposed Policies? Science of the Total Environment, 472, 282–288.

European Environment Agency. (1998). Europe's Environment: The Second Assessment.

Jacobs, M. (1991). The Green Economy. London: Pluto Press.

Meadows, D.H., Meadows, D.L., Randers, J. and Behrens, W.W. (1972). The Limits to Growth. New York: Universe Books.

Pearce, D.W., ed. (1993). Blueprint 3: Measuring Sustainable Development. London: Earthscan.

Roy, A. and Goll, I. (2014). Predictors of Various Facets of Sustainability of Nations: The Role of Cultural and Economic Factors. International Business Review, 23, 849–861.

Scrase, T. (1999, August). The Judicial Review of LPA Decisions: Taking Stock. Journal of Planning and Environmental Law, 29, 679–690.

Tukker, A. (2013). Knowledge Collaboration and Learning by Aligning Global Sustainability Programs: Reflections in the Context of Rio+20. Journal of Cleaner Production, 48, 272–279.

Vergragt, P., Akenji, L. and Dewick, P. (2014). Sustainable Production, Consumption and Livelihoods: Global and Regional Research Perspectives. Journal of Cleaner Production, 63, 1–12.

World Commission on Environment and Development. (1987). Our Common Future. Oxford: Oxford University Press.

9 Conclusion

Begum Sertyesilisik, Ahmed Al-Shamma'a, Amr Sourani, Anupa
Manewa, Bernd Kochendoerfer, Margarete Roigk, Basak Guçyeter,
H. Murat Gunaydin, Tofigh Tabesh, Laurence Brady, David Phipps,
Derek King and Jack Rostron

This book has examined the technical, managerial, legal and economic aspects of
the sustainable built environment. Managerial aspects have been analysed focusing on sustainable procurement, cost modelling for sustainability and sustainable
building process based on three main phases: pre-construction, construction and
post-construction. Technical aspects have been investigated with respect to sustainable buildings (i.e. building assessment tools, characteristics of sustainable
buildings and passive design), including low- and zero-carbon technologies in
buildings, as well as water efficiency and related utilities in sustainable buildings.
Legal aspects have been investigated focusing on the environmental law for the
built environment and environmental impact assessment. All these aspects, like
puzzle pieces, are important ingredients for the sustainable built environment. In
this chapter, we make some concluding remarks for each of the chapters and then
possible future trends for all these aspects are suggested.

Sustainable procurement: Chapter 2 emphasised the importance of sustainable procurement within the construction context and explained that "sustainable procurement" is one of the significant criteria in the procurement channel,
which leads towards "sustainable construction" and "sustainable developments".
Different countries pay attention to different areas of sustainable procurement
(such as health and safety, diversity etc.) aiming to achieve sustainability goals.
Sustainability principles can be integrated into a project brief and contract specifications. Multi-criteria decision-making techniques can be used to make informed
decisions that consider sustainability criteria. The contractors should be assigned
or appointed considering value-based rather than price-based selection principles
so that sustainability performance of the procurement process is enhanced. At
the tender stage, contractors may be encouraged to identify sustainable solutions
that can result in life-cycle savings, which can then be shared, as an incentive,
through certain mechanisms. Other enablers that could enhance the development of "sustainability-oriented" procurement are related to issues of knowledge and perception, or could be political and regulative, financial, instrumental,
logistical issues, organisational and management and strategic issues.

Cost modelling for sustainability: In Chapter 3, cost modelling for sustainability was discussed in an array of methods and tools, which are directly related
to a multi-variable decision-making method and life-cycle cost assessment. Cost
modelling for sustainability is a concept that cannot be evaluated independently
from the environmental and social aspects of sustainable built environment. In
the conventional project management process, economic performance is assessed
through time, cost and quality. Due to the rise in environmental concerns, however, life-cycle cost assessment has become a useful tool in building evaluation

procedures. It is examined as a methodology that has the potential to evaluate financial sustainability measures in the built environment, estimating environmental impacts and energy consumption patterns. To assist investment decisions within a holistic perspective, Chapter 3 described economic interactions and focused on the costs that are incurred during the lifespan of a building. Chapter 3 emphasised life-cycle cost assessment as a methodology that can significantly contribute to sustaining resources, both physically and financially, as it can contribute to an enhanced productivity in sustainable construction. Simultaneous optimisation of investments, resources and environmental impact enables a positive impact on a sustainable future.

Sustainable building process: Chapter 4 focused on enhancing the sustainability performance of construction project management in three main phases, namely: pre-construction, construction and post-construction. The environmental concerns and sustainability aspects need to be considered throughout these phases. The tendering phase needs to result in the assignment or appointment of a contractor, subcontractor or design company with expertise and experience in the field of sustainable construction and sustainability requirements. The language of the contract and its documents should be precise, enabling the rights and obligations of the contracting parties to be clearly laid out so that the need for re-work is reduced. The involvement of the contractor in design activities is recommended for reducing the need for re-work that can cause an increase in the amount of wasted materials. The design should envisage the whole life-cycle period of the project. Material selection, specification and procurement activities should cover environment friendly procedures, avoiding hazardous chemicals and so on. The requirements of the building assessment tool, which is referred to in the contract, should be fulfilled. Throughout the mobilisation, construction and demobilisation phases, effective site waste management should be carried out. Assigned or appointed subcontractors should be responsible for their sustainability performance on site and they should be trained in how to enhance this. Sustainability performance of construction project management can be enhanced through benchmarking management principles and techniques used in other industries. Lean management principles can be adapted to the construction industry and to all phases of the construction project management, under the name of lean construction, to reduce waste and to create value.

Sustainable buildings: Living in harmony with nature is key for future generations to survive and for sustaining nature. Chapter 5 focused on sustainable buildings. It covered building assessment tools and key technical aspects of the sustainable buildings. It revealed the importance of selecting sustainable building materials, of proper design of the building envelope and the importance of benefitting from passive systems for energy efficiency.

Low- and zero-carbon technologies in buildings: Chapter 6 provided an insight into the use and development of low-carbon technologies in buildings. Obviously, the energy requirements of a building are greatly influenced by the particular climate of its location. However, growing infrastructure needs in much of the world require that technology will play an important part in maintaining modern societies. Modern lifestyles mean that man-made infrastructures will be vital necessities and it is now recognised that construction professionals must play an important role in managing the infrastructure sustainably. The holistic interpretation of sustainability, in which the implications of how development will affect people, their environment and well-being, as well as their economies, gives an international aspect to the consideration of low-carbon and renewable technologies. Increased atmospheric carbon can present risks to all countries. Different renewable sources and opportunities exist in different areas of the world and therefore solutions may differ from place to place. A design philosophy in which architects and engineers approach problems co-operatively is increasingly being adopted. Although in some cases the team approach has developed because of national regulations, many construction professionals now recognise how passive and active solutions can both contribute to low-carbon designs.

Sustainability in utilities: water efficient sustainable buildings: As Chapter 7 showed, the drive for domestic water efficiency is surprisingly complex with a range of both technical and socio-economic problems to be addressed. Moreover, the problem is huge in scale so any initiative must be cost-effective and readily applied to large segments of the building stock. Whilst the installation of water efficient devices can be enforced in a new build, this is only a small proportion of the stock. Decisions based on both economic and environmental factors will have to be made on the scale and pace of retrofit in the existing stock of buildings. There will be a constant need to update and improve codes of practice. Some advantage may be gained from enforcing water efficiency measures under a certification scheme before properties can be sold. However, technical solutions are only effective if consumer lifestyle changes can be made to support their use. Purely economic restrictions based on metering and an associated price regime can be helpful, but this must be adopted sensitively to avoid water poverty. The key will be an education policy that encourages adoption and retention of a positive attitude to water saving and so results in water efficient personal habits. There is a considerable tension between the consumer lifestyle and the drive towards water efficiency. Finally, as heating water accounts for a large part of domestic energy consumption, water saving should be linked in the consumer's mind to energy saving, which currently has a much higher profile.

Environmental law for the built environment and environmental impact assessment: Chapter 8 emphasised the need for international collaboration and effective local actions so that the world habitat can be sustained and restored. The concepts of sustainability and development of environmental impact assessment techniques are closely interrelated. The two topics are not only intertwined as concepts, but the geographical distribution of environmental pollutants requires transnational administrative protocols and wide legal jurisdictions. In recent times this has resulted in the creation of bodies such as the European Environment Agency. The evaluation of environmental impact, in terms of the breadth and depth of topics covered, is clearly related to the perceived level of environmental impact damage. Some traditional practitioners will argue that the development of alternative designs already intrinsically involved evaluation of potential hazardous impacts on the environment. Whilst there is clearly a need for the sustainability of development to be considered in the formal context of statutory environmental impact, the process has clearly added to the complexity and therefore cost of gaining planning permission. Some would argue that it simply adds more delay and expense to the process of securing planning permission.

This book has analysed the sustainable built environment with respect to technical, managerial, legal and economic aspects. These aspects are supported by social aspects as the human factor plays an important role in the successful transformation of the built environment into a sustainable one as we all influence the demand structure and consumption patterns. The increased awareness of sustainable development will encourage and enforce the construction industry to reduce its environmental footprint resulting in the enhanced sustainability performance of the built environment and of construction project management. Policies, education and training institutes, as well as media, can all act as main drivers towards sustainable development as they can raise our awareness with respect to the deterioration of the world habitat and our role in sustaining the environment.

The possible future trends towards an improvement in the sustainability performance of the built environment and construction project management with respect to the technical, managerial, legal and economic aspects we have been discussing can be briefly summarised as follows:

- *Technical aspects*: We can anticipate more innovation and IT-intensive work, enabling dematerialisation and consequently the enhanced sustainability performance of the built environment. A new trend in the near future is expected to be the shift from sustainable buildings towards regenerative buildings, which aim for net positive impact to the environment. This new concept, which is in its infancy, will require new technologies and new materials. As climate change continues to occur, causing disasters with increased frequency and magnitude, the design and technical aspects of the built environment need to become more disaster resilient.

- *Managerial aspects*: The expected changes in technical aspects should lead the management process to be more iterative and interdisciplinary, requiring multi-talented project managers and staff. The changes should lead to the emergence of new managerial concepts and solutions as well as new building assessment tools. Management principles used in other industries can be adapted to the construction industry where appropriate. Managerial aspects can be supported more through innovation and IT-enhanced coordination and collaboration among contracting parties and stakeholders. Higher Education institutes need to adapt themselves and their curriculum to the changing requirements of the construction industry so that they can lead innovation and educate the professionals who will need to be able to deal with the challenges of the industry's future.

- *Legal aspects*: Changes in technical and managerial aspects can lead to changes in legal aspects, which can in turn provide solutions to the changing needs of the industry. Accordingly, new environmental laws and regulations can be established. Furthermore, as dealing with climate change and sustaining the world habitat requires immediate international collaborative action, new international treaties and directives can be launched enabling local laws and regulations to be synchronised to enhance the environmental sustainability performance of all countries.

- *Economic aspects*: Imposing footprint taxes and adding the cost of damage to the natural environment to the production process are on the world agenda. Such changes can affect cost calculations and the accounting system leading to the emergence of new concepts and changes in competitive advantages of companies.

Index